E-MANUFACTURING: BUSINESS PARADIGMS AND SUPPORTING TECHNOLOGIES

Editorial Board

E-MANUFACTURING: BUSINESS PARADIGMS AND SUPPORTING TECHNOLOGIES

18th International Conference on CAD/CAM Robotics and Factories of the Future (CARs&FOF)
July 2002, Porto, Portugal

Edited by

João José Pinto Ferreira
Faculdade de Engenharia da Universidade do Porto
Instituto de Engenharia Sistemas e Computadores do Porto
Portugal

SPRINGER SCIENCE+BUSINESS MEDIA, LLC

Library of Congress Cataloging-in-Publication Data

A C.I.P. Catalogue record for this book is available from the Library of Congress.

E-Manufacturing: Business Paradigms and Supporting Technologies
Edited by João José Pinto Ferreira
ISBN 978-1-4613-4730-9 ISBN 978-1-4419-8945-1 (eBook)
DOI 10.1007/978-1-4419-8945-1

Printed on acid-free paper.

CONTENTS

CO-SPONSORS

INESC PORTO

INSTITUTO DE ENGENHARIA DE SISTEMAS E COMPUTADORES DO PORTO

Universidade do Porto

FEUP Faculdade de Engenharia

FCT Fundação para a Ciência e a Tecnologia

MINISTÉRIO DA CIÊNCIA E DO ENSINO SUPERIOR Portugal

Apoio do Programa Operacional da Ciência, Tecnologia, Inovação do Quadro Comunitário de Apoio III

18th INTERNATIONAL CONFERENCE ON CAD/CAM, ROBOTICS AND FACTORIES OF THE FUTURE

Patrons
Carlos Costa
Marques dos Santos
Pedro Guedes Oliveira

Conference Chairman
J. J. Pinto Ferreira

Co-Chairman
L. M. Camarinha-Matos, G. Bright, H. Bera, José M. Mendonça, R. Gill, Y. Kakad

Local Organizing Committee
Américo de Azevedo
Ângelo Martins
António Lucas Soares
Diogo Ferreira
José Carlos Caldeira
Luís Maia Carneiro
Sónia Pinto

International Program Committee
A. Azevedo, PT
A. Paulo Moreira, PT
A. Pessoa Magalhães, PT
A.L.Soares, PT
Abdelhakim Artiba, BE
Adil Baykasoglu, TR
Adriano Carvalho, PT
Alcibíades Guedes, PT
Alexandra Klen, BR
Ângelo Martins, PR
António Moreira, US
Arturo Molina, MX
AS White, UK
Ben Rodriguez, BE
Bernhard Koelmel, DE
Camarinha-Matos, PT
Carlos Couto, PT
Carlos Ramos, PT
David Chen, FR
ET Smerdon, US
Eugénio Oliveira, PT
F. Restivo, PT
F. Vernadat, FR
Felipe Martins Muller, BR
Fernando L. Pereira, PT
Florent Frederix, BE
G Bright, ZA
G. Putnik, PT
Guenter Schmidt, DE
Guy Doumeingts, FR
H. Afsarmanesh , NL
Howard Richards, UK
J. Falcão Cunha, PT
J. Fernando Oliveira, PT
J. Rubinovitz, IL
J.J.Pinto-Ferreira, PT
J.M.Mendonça, PT
J.P.Sousa, PT
Jan Goossenaerts, NL
Jean Pol Piquard, PT
João Monteiro, PT
Joerg Pirron, DE
John J. Mills, US
Jorge Gasós, BE
José Barata, PT
José Faria, PT
José Paulo Santos, PT
K.R. Caskey, NL
Kurt Kosanke, DE
Laszlo Nemes, AU
Luc de Ridder, BE
M Carvalho, BR
Mário Lima, PT
Markus Rabe, DE
Marques dos Santos, PT
Martin Ollus, FI
Norberto Pires, PT
P. Bernus, AU
P. Veríssimo, PT
Paulo Miyagi, BR
Peter Kopacek, AT
R. Gill, UK
R. Freudenbreg, DE
Ricardo Rabelo, BR
Sá da Costa, PT
Sam Bansal, SG
Shahin Rahimifard, UK
Sikir K Banerjee, UK
Sílvio Carmo Silva, PT
Stephen T Newman, UK
Svetan Ratchev, UK
T. Goletz, DE
T Ito, JP
T King, UP
Ted Goranson, US
Turkay Dereli, TR
Uwe Kirchoff, DE
V Patri, HK
Volker Stich, DE

FOREWORD

The International Conference on CAD/CAM Robotics & Factories of the Future (CARs&FOF) has been organised in several locations around the world for almost two decades. Under the topic "E-Manufacturing, Advances in business paradigms and supporting technologies", this event took place in July 2002 in Porto, Portugal, a joint organisation of:

INESC Porto
Faculty of Engineering of the University of Porto
International Society for Production Enhancement

and with the support of the Portuguese Foundation for Science and Technology.

This book includes a selection of the papers presented in CARs&FOF'2002 as well as invited papers reflecting the vitality of the discussions held in the conference plenary sessions. This is the consequence of extensive teamwork, combining the generous collaboration of the invited editorial board members that jointly selected the papers herein published. These acknowledgements should be further extended to the Conference Organising Committee, Conference Co-chairman, Conference Secretariat and finally to the support given by Mrs. Sónia Pinto and Mrs. Marta Oliveira in the final editing of this book.

The Editor

João José Pinto Ferreira
July 2003

EDITORIAL BOARD FOREWORD

E-Manufacturing implementation and related business practices demands an ever increasing knowledge about enabling technologies. Moreover, the digital and wireless world will surely trigger new business practices. This conference aims therefore at bringing together the business and technology research worlds into a catalyst forum for new business models and technology opportunities. In this context we have the objective of bridging the gap between technical and business discussions, usually developing in widespread forums.

E-Business is a very significant economic and social paradigm and is now building on the convergence of several technologies. Under this topic, and concentrating our efforts in the E-Manufacturing area, we aim at highlighting strategies, methods, the demand for deployment of new business models and for intra- and extended-enterprise business processes.

This book opens with a set of interesting selections from invited authors, covering perspectives such as concurrent engineering in product and process design, the tools needed to deal with people, relationships and networks, enterprise networking in Europe. This section closes with business and innovation topics, handling issues such as knowledge, innovation and investment, and joint ventures for innovation and competitiveness. The remaining parts of the book tackle the following e-manufacturing issues: advanced logistics, mechatronics, manufacturing systems integration and supporting technologies.

The Editoral board

Américo Lopes Azevedo
Ângelo Martins
António Lucas Soares
Robert J. Graves

PART ONE
INVITED PAPERS

PERSPECTIVE ON CONCURRENT ENGINEERING IN PRODUCT AND PROCESS DESIGN

Dr. Robert J. Graves
John H. Krehbiel Professor of Emerging Technologies
Thayer School of Engineering
Dartmouth College
8000 Cummings Hall
Hanover, New Hampshire 03755-8000
USA
robert.j. graves@dartmouth.edu

Abstract: *There have been many advances in the processes of product design and development over the last 25 years. Some of the advances have been in technology support tools and some have been in the product realization processes themselves as recognition of the technology capabilities develops and matures and incorporates the business practices that guide the development practices. These advances are briefly described in this paper as we observe the maturation of CAD/CAM technologies, the development and improvement of Design for Manufacturing technologies, the change in business practices from vertically integrated firms to distributed supply chain-oriented enterprises, and the increasing use of network-based approaches in E-engineering.*

Keywords: CAD/CAM, Product Development Processes , E-engineering

INTRODUCTION:

The engineering of products is often a complicated and technical process that occurs under pressures of time constraints, cost constraints, and quality issues. One way in which to view design is as a series of activities by which the information about an object is created and recorded. It is a creative process and one which changes the state of knowledge about an object. As engineering design progresses, the amount of information available about the designed object increases and becomes more detailed in its nature. It is no wonder then, that engineering design relies so heavily on good quality information and the associated technologies that support its access and use in design of the object.

Engineering design is an integral part of a product realization process. It is joined by the determination of customers' needs and the relationship of these needs to company strategies and products, development of marketing concepts, development of engineering specifications, then the design of both the product and the tools and the processes by which it will be made and assembled, followed by determining the approaches to distribution, sales, repair and disposal. The focus of this paper is on the engineering design process and its integration with fabrication and manufacturing, hence leaving out the market concept development and specifications development as well as the later aspects of product realization dealing with distribution, etc.

There have been many advances in the processes of product design and development over the last several decades. Some of the advances have been in technology support tools. Others have been in the product realization processes themselves as recognition of the technology capabilities develops and matures and begins to incorporate the business practices that guide the development practices. Such advances are considered in this paper in the context of engineering design as a creative process involving information retrieval and use as well as information generation. These advances are briefly described in the following as we observe the maturation of CAD/CAM technologies, the development and improvement of Design for Manufacturing technologies, the change in business practices from vertically integrated firms to distributed supply chain-oriented enterprises, and the increasing use of network-based approaches in E-engineering.

PRODUCT DEVELOPMENT PROCESSES AND TOOLS:

Traditional product development tools relied on drafting boards and later, in the 1970s, the use of Computer-aided design (CAD) tools. While the original CAD tools were little more than automated drafting technologies, they have matured and become more integrated with Computer-aided Manufacturing (CAM) technologies. Advances through the 1980s and beyond were both technological and process or procedural in nature.

1980s

The common business environment of the 1980s was one of vertically integrated firms with internal fabrication and assembly shops. Product development engineers used drafting tools and CAD tools in their functions of product design. When the need arose, these engineers would consult with manufacturing engineers to gain knowledge of the issues in fabrication and assembly associated with their product designs. One might view this knowledge as readily at hand, in that the fabrication and assembly shops were either close by or easily accessed through internal corporate processes.

The common product realization process, as depicted in Figure 1, might also be characterized as the "over-the-wall" practice where product design was the initial phase and little discourse with fabrication and assembly took place. When design was completed, the design was then handed to manufacturing for review and comment and this may or may not have led to design adjustment. In a similar manner, manufacturing would indicate to purchasing what was needed from a material perspective and purchasing, in turn, would interact with suppliers to order necessary purchased parts. The limitations of the over-the-wall practice then become clear in that there is limited interaction between functions or phases of the process to help support the support the design development in ways that aided cost efficiencies, improved manufacturing of internal parts and improved assembly ease for the product.

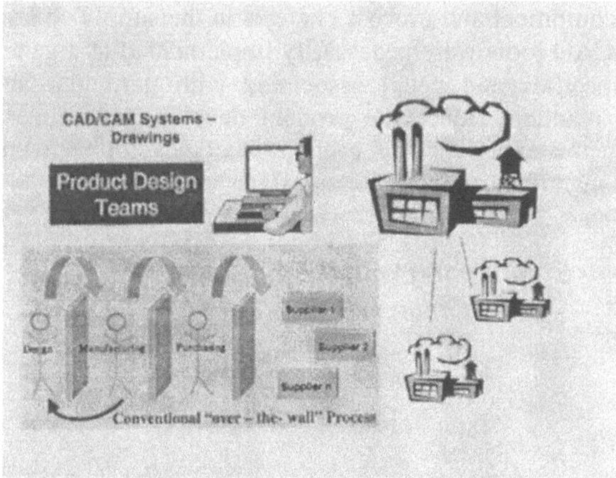

Figure 1: Product Development in 1980s

As a consequence, the 1980s began to see the development of design-aid tools which intentionally captured manufacturing and assembly knowledge into evaluative technologies for use by design engineers. The advent of Design for Assembly analysis and various Design for Manufacturing (e.g. Injection Molding, Die Casting, and Stamping) design aids heralded the era of improved design evaluation at the early phases of design by the consideration of fabrication and assembly consequences of design choices. Some of these design aids were analytical and some were qualitative in nature, but they supported what was increasing being called "concurrent engineering" as the 1980s gave way to the 1990s.

1990s

Business practices began to change in the 1990s and these changes affected the product realization processes. Firms which had been vertically integrated began increasingly to "outsource" not only purchased parts for their products but also fabrication and perhaps assembly functions. This phenomenon led to the development of enterprises composed of "original equipment manufacturers" or OEMs and their associated supply chain members.

This migration of fabrication and assembly functions out of the former umbrella of the vertically integrated company into separate corporate entities able to compete on their own, led to changes in product development. The knowledge of fabrication and assembly that had been shown to be useful in product design now was less accessible to the design engineers with the increased distance of geographical separation and organizational boundaries. Product realization processes, as seen in Figure 2, now had to rely upon exchange of information through communication devices (e.g. fax, phone) or perhaps surface mail with contracted suppliers to gain design review knowledge. This increased difficulties due to communications and generated delays (and often increased costs) in the processes of product realization. The tools developed in the 1980s to support design through fabrication and assembly analysis often became outdated fell as their maintenance was unable to keep pace with equipment and process changes in the supply chain. Those that were incorporated in CAD tools were necessarily implemented at a general level and lost the utility of specifics and detail associated with particular supply chain firm technologies or practice. Hence the product development process tended to rely more heavily on the experience of each product development engineer to assure consideration of manufacturing issues in design.

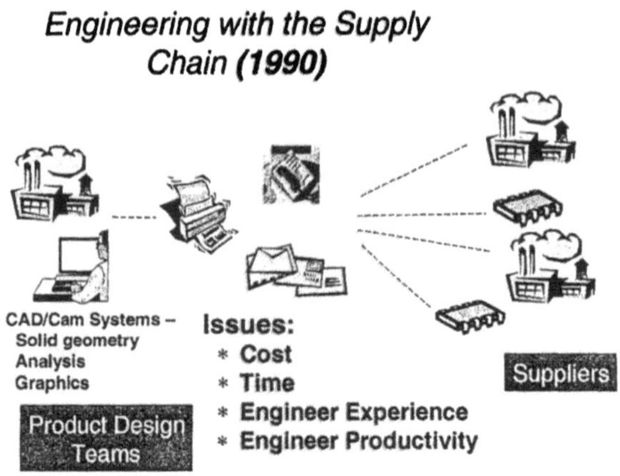

Figure 2: Engineering with the Supply Chain in 1990s

2000

A critical technology maturing through the 1990s was the Internet. Widely distributed networked computers had heretofore been available to the defense industry, to academics and to selected computer companies, but not generally accessible to the broad industry and service company base. This technology has yet to reach full maturation, but the applications and support for product realization practice were appearing in the late 1990s and expanding into the early 2000s. Figure 3 depicts a product realization practice in the electronics industry where printed circuit board assemblies are designed relying upon board fabrication and aboard assembly service suppliers as well as component suppliers which often reside in the enterprise supply chain rather than captive under the OEM organization. Since product design is a driver, this might be termed a "Design Chain" where knowledge and its exchange is more important than providing material as is the case in a Supply Chain.

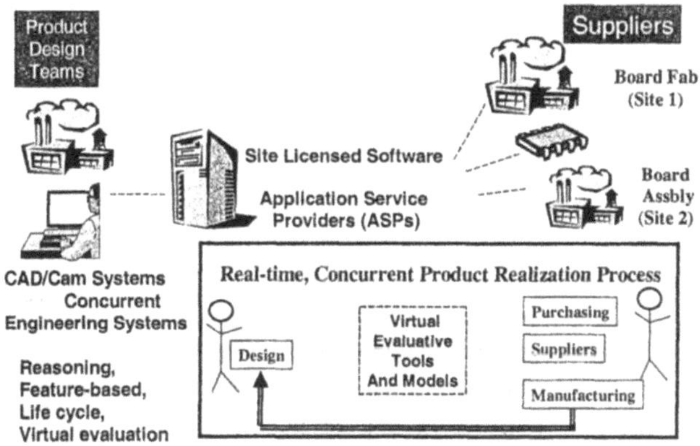

Figure 3: E-engineering with the Supply/Design Chain

The integration of advances in CAD/CAM technologies, such as solid geometry, feature-based reasoning, and life cycle analyses together with network-based technologies used to access knowledge bases and databases provided new and powerful support to product realization. Design engineers could develop automated and real-time review of their designs through network tools where the maintenance of the knowledge bases and databases were distributed among the respective design chain entities. This provides up-to-date evaluation, based on timely information and accomplished nearly instantaneously, allowing for design improvement with less delay and cost in the process. An additional feature of this practice is the decreased reliance on individual design engineer experience in terms of manufacturing knowledge, since the knowledge is maintained by the supply chain entities and accessed through the technology.

Concurrent Engineering:

The shifting nature of concurrent engineering thus becomes clear. In the 1980s, it is characterized by verbal design reviews usually occurring after design was complete. The knowledge used in design review was largely experience-based (often accessed locally) or directed by policies of standardization within the vertically integrated firm. CE in the 1990s transitioned to use of centralized analysis tools, often implemented as specific software tools in a separate so-called "back-end" analysis to the CAD software, which were built upon knowledge and data from manufacturing. The success of DFA and DFM technologies in the 1990s came from their more widespread use in product design within standardized design environments of vertically integrated firms but was subsequently limited by the nature of business practices shifting to geographical distribution of fabrication and manufacturing services. Now, in the 2000s, we see concurrent engineering becoming e-engineering in environments of networked design chains for the enterprise. The e-engineering tools include imbedded and decentralized models for design analysis and cost and time-to-market estimation. Distributed databases and knowledge bases incorporate distributed responsibility for their maintenance and assure more timely information to the product realization process, with the consequent result of improved product design in less time as well as reductions in product realization times overall.

E-engineering and the Design Chain in the Future:

The technologies and processes of engineering design are not standing still. Manufacturing processes are also changing, becoming more expensive, more technology driven and more precise, and perhaps more rigid in nature. In the 1980s and earlier, it appeared that processes were designed to support the product design as the "over-the-wall" syndrome was well entrenched. Increasingly through the 1990s and into 2000s, the products are being designed to fit or take advantage of particular processes. This change, of course, places increasing reliance on data and knowledge about specific manufacturing processes to be used in the engineering design process. Ideally, such knowledge is incorporated in an environment where it is readily renewed or maintained along with the pace of process /material technology changes and thus decays in its quality and timeliness do not occur.

The trends toward increased global sourcing of parts and manufacturing services are likely to continue. These trends are driven by competitive pressures for faster, better and more cost-effective product solutions for customer needs. Increasing pressure for mass customization is also affecting product design and product design processes.

From a technology perspective, the advent of more robust technologies which are inclusive of customer involvement, with efficient use of distributed knowledge bases and databases is near. The degree to which such technologies can directly link to a CAD/CAM technology, thus more fully integrating the tool suites of the design process, is still unclear due to such issues as continued reliance on non-networked CAD/CAM software, format and characterization issues for information, and development of scalable approaches to support customer activity directly in the design process itself.

The engineering of products will continue to be a complicated and technical process that occurs under pressures of time constraints, cost constraints, and quality issues. It will continue to be a creative process and one which changes the state of knowledge about an object. The focus on information technologies to support the design process with better information and better access to knowledge useful in design is where future technology advances will derive.

We noted earlier that engineering design is an integral part of a product realization process, and we will expect the future to see extensions of integrated tools to be used in design that include functions in both upstream and downstream directions. Such technologies would therefore help the design engineer to incorporate considerations other than functional design into the engineering design of future products.

ACKNOWLEDGEMENTS:

The EAMRI at Rensselaer was established under sponsorship of the U. S. National Science Foundation (DMI-9320955). The EAMRI's work is also supported under DMI-0075524 as well as a recent second grant from the National Science Foundation (DMI-0121902) under its Scalable Enterprise Systems Initiative. EAMRI work is also partially supported by the EAMRI's collaborating companies including CISCO Systems, Lucent Technologies, Hughes Missile Systems company (now part of Raytheon), Texas Instruments Defense Systems (now part of Raytheon), Pitney Bowes, Vermont Circuits, Universal Instruments, MENTOR Graphics and SmartLynx.

E-MANUFACTURING: THE TOOLS WE NEED TO DEAL WITH RELATIONSHIPS AND NETWORKS

António Lucas Soares
INESC Porto and Faculty of Engeneering of University of Porto
Rua Dr. Roberto Frias, s/n 4200-465 Porto, Portugal
antonio.lucas.soares@fe.up.pt

This paper proposes a social networking perspective on the e-manufacturing area. A short justification of the social network perspective as applied to manufacturing organizations is made, focusing on methods and theoretical approaches. Then, it is speculated how social networking can be a paradigm for the management of networks of manufacturing enterprises. The management of relationships (fostering collaboration and cooperation) and their support through new information systems architectures are the key issues. Finally, some research directions are discussed, focusing on new engineering paradigms.

Keywords: Networks of organizations, social networks, networking, hyper-management, management information systems

1. INTRODUCTION

"E-manufacturing is concerned with the use of the Internet and e-business technologies in manufacturing industries. It covers all aspects of manufacturing - sales, marketing, customer service, new product development, procurement, supplier relationships, logistics, manufacturing, strategy development and so on. The Internet also affects products as well since it is possible to use Internet technologies to add new product functions and to provide new services." (Cheshire Henbury, 2001). This definition of e-manufacturing places this concept in the general context of inter -organizational collaboration and cooperation, or what we can call a "networking context".
In the last years there has been an interest on the networking concept as applied to management and technology studies. The general theoretical framework of the networked society (Castells, 1996) strongly supported by a set of comprehensive empirical studies, settled up a new analysis perspective triggering several research programs developing and using network-centric theories and methodologies. "Networks are the fundamental stuff of which new organizations are and will be made" says Castells adding further that the network will become the basic unit of economic organization. The networking phenomena are not exclusive of the area of economics but spread to all areas of social activity, from solidarity to culture. It is within this context that e-manufacturing is nowadays developing. Today we find within the manufacturing area supplier networks, producer networks, customer

networks, coalitions for standards, technology cooperation networks, research networks, collaborative heterogeneous networks, etc. which become more and more complex, and more and more demanding in terms of coordination and general management. These requirements call for new perspectives and paradigms in engineering e-manufacturing systems.

2. NETWORKING: THE CONTEXT FOR E-MANUFACTURING

2.1 From chains to networks

In the manufacturing world, supply chains have been the most spread type of network of organizations. In the last years, the concept of supply chain has gone beyond the focus on logistic processes and is nowadays advancing steadily towards the integration of key business processes, from suppliers to final customers, encompassing products, services and information (Lambert and Coper, 2000). More and more, companies interact with the clear purpose of cooperation. Some authors call this *networking* (Magnusson and Nilsson, 2003; Triekenens, 2002).
Generalizing, we can consider the organizational network concept defined as a physically decentralized social network made up of individuals who form a community but are not members of the same formal organization (Howard, 2002). These organizational networks are called for example 'communities of practice' in sociology (Bijker *et al.*, 1987) or 'knowledge networks' in management (Podolny and Page, 1998; Uzzi, 1996). We can identify organizational networks in each of the specific types of networks enumerated above.
For example within the manufacturing area, knowledge networks are becoming particularly important. One key factor in such networks is the ability to integrate the knowledge managed by each of the organizations. This happens in supply chains, where the level of knowledge integration is low, in business networks, built upon a central actor (hub-firm) that identifies a business opportunity and creates a network to satisfy that need, and have a medium knowledge integration level, and in research networks where knowledge creation is the main goal and a high level of knowledge integration is achieved (Magnusson and Nilsson, 2003). A range of methodological approaches have been used to study these social phenomena (depending of course on the aim of the study), but every approach is built upon a set of models used to explain or just describe relevant aspects of those phenomena (Howard, 2002).
Networks can be seen as an evolution of chains in the sense that they possess a stronger social structure, which helps to generate trust, stimulate networked learning and to internalize external issues such as technologies or standards (Jonkers, and *et al.* 2001).

2.2 Studying industrial collaboration and cooperation as networking

The objects of study of networks and chains encompass, besides the actors and their relationships, the patterns of activities creating and transferring value objects. These patterns are referred as the network business processes and the pattern of relationships associated as network structures (Jonkers, *et al.* 2001). Nevertheless, there is a complex inter-relation between structure and processes in networks of organizations (Barbedo *et al.*, 2003).

The objects of study of networks and chains can also be analised from three main viewpoints: (1) socio-economic, where networks and chains are characterized by the creation and exchange of value through specific functions and mechanisms, by actors having common or conflictual interests; (2) informational, where networks and chains are viewed as information processing systems required for control and management; (3) technical, where networks and chains are viewed as distributed production systems characterized by technical and logistics transformations.

Particularly on business and technology studies, and in smart organizations and collaborative networks development, network-centric approaches are being extensively used. Two of these approaches are currently been given some focus (Lamb *et al.*, 2001): social network analysis and actor-network theory.

Social network analysis builds on a long tradition of social-structural perspectives. Nevertheless, there was recently a resurgence of this approach (Burt, 2000; Zack, 2000; Lamb *et al.*, 2001). In networked or virtual organizations this approach is well suited as social structure is not bound to the formal organization and the analysis may apply naturally to inter-organizational relationships. For example, the work of Wellman et al. (1996) on cooperative work and telework builds directly on social network concepts. Gitell's work (2000) also highlights the role of social networks and the use of various enabling technologies (including ICTs) to support forms of relational coordination (Lamb *et al.*, 2001).

Another new research direction also draws on sociological theory to develop a better understanding of social interaction and information technology and systems. Latour's (1987) actor-network theory (ANT) combines the broad-scale thinking of the SST tradition with new conceptualizations that raise technologies (such as computers and networks) to an equal status with human actors (Lamb and Kling, 2001). This perspective explores the intricate interrelationships that develop between people and the technologies they employ to interact with other individuals, organizations and institutions within complex, interconnected networks (Walsham, 1997).

3. SOCIAL NETWORKING AS A MANAGEMENT PARADIGM FOR E-MANUFACTURING

3.1 Managing as "managing relationships"

The theoretical outline of the previous section can be applied in the exploitation of new concepts for management, especially the management of networks of small and medium enterprises. For example, we can take a structural approach to the modelling of an SME network by considering two key concepts: social networking and hypertext. Here social networking refers to a range of approaches both from sociology and organization theory (as seen before), where the organization and their members are viewed as nodes in networks. Hypertext is a concept that underpins today's forms of interaction in the internet and information technology in general. From these concepts we invent the concept of hyper-management with the goal of forming, in some extent, a unifying concept. Ideally this would mean to consider both the local and the global, the detailed and the aggregate, the technical and the social, the structure and the behaviour.

The management of an enterprise, or a network of enterprises, can be viewed as the management of relationships and interactions between the different actors, directly and indirectly involved in the activities. Relationships involve operations, processes, resources, knowledge, social interaction, trust, power, etc. Relationships can be related with other relationships. For example, an operational relationship between a procurement manager and one of its suppliers is related (is influenced) by his relationship with the other suppliers, by the relationship of the supplier with other clients, etc. This can be considered an evolution of the concept of meta-management (Mowshowitz, 1997) in virtual organizations, going from the management of distributed tasks linked to some previously agreed overall goals, to the management of a web of relationships linked not only to a complex web of overall goals but also to individual goals and expectations.

The way a manager (which can be any node in a network) moves from one subject or issue to another strongly depends on the linkages (relationships) he actually perceives and on his goals and objectives at the moment. Therefore, there can be several levels of relationships. For example, depending on the level of aggregation of the actors (e.g. nodes in a collaborative network) we can explore the relationships between companies, between teams, between roles, etc.

A holistic management view also requires managing the relationships and interactions of individuals and groups with technology, by adequately positioning the technological artifacts, such as software agents in information systems, in the social context of human action. In fact, today, information technology and systems have a so strong influence in human actors' relationships that it is interesting to conceptualize them as social actors on their own.

Through a "multi-perspective model" managed by a computer, each person or group is expected to be able to fully explore webs of relationships that are important in the scope of his responsibilities. Relationships can be analysed, "designed" and managed at different levels and from several perspectives. Levels and perspectives are intertwined, linking the network of social actors in a hyper-web of relationships.

3.2 Information systems for managing relationships

The functional/hierarchical foundational concepts of past management information systems are not compatible with the requirements of hyper-management. Rigid whole-part relationships (hierarchies) and functional dependencies are not appropriate to manage the complexity of today's webs of relationships in business networks. The need to freely navigate between management issues such as operations, collaborations, negotiations, plans, etc., call for new systems' concepts and new information technology models away from the simple client/server architecture. Intelligent agents technology, peer-to-peer and grid computing are emerging in the business community as promising business information systems platforms. Nevertheless, the breakdown with the old models of management information systems has not yet occurred. Nowadays, new management concepts can only emerge in a complex interaction with new information systems concepts.

A new concept for management information systems in business networks should be based on three issues: networked collaboration support, networked decision support and networked knowledge management. This requires multi-perspective (inter) enterprise models, including knowledge models and decision making models appropriate to enable the best management decisions by every individual and group. This model is inherently distributed and thus agent technology, peer-to-peer and grid computing naturally fit the implementation requirements of such model.

A new reference architecture for management information systems should cope with two characteristics of future collaborative networks: heterogeneity and complexity. For that, highly flexible information technology infra-structures are needed. Intelligent agents, peer-to-peer and grid computing are promising technologies to answer these requirements. Agent based architectures provide a platform for the integration of heterogeneous services such as order distribution support in logistics management or skill capital optimization in human resources management. Software agents and in particular intelligent agents technology provide both the way to implement flexible and sophisticated business functionalities and the way to encapsulate the existing (legacy) ones in a way that reflects the networked nature of hyper-management. Peer-to-peer and grid computing provide a much more flexible and powerful way of processing management data at several levels than current data warehouse bases architectures.

4. DISCUSSION

The ideas presented in the previous section point one direction of research work that fits the needs of advances in e-manufacturing, answering to the 4 levels of complexity identified by (Moller and Halinen 1999): (1) industries as networks - vision of the entworks; (2) industries in the network - management of the network; (3) relationships portfolio - portfolio management; and (4) exchange of relationships - management of relationships.

More specific research directions would include to consider methods for the design of inter-organisational relationships including explicit models of the social-actors involved in such co-operation processes. This leads to a new engineering paradigm, inter-disciplinary by nature, where the network is both the development object and the means to undertake that development. For example, conceptual and computer based tools to be used in modelling for the analysis and management of inter-organizational processes hybrid modelling: qualitative / quantitative, technological / sociological, conceptual / mathematical; reference modelling, "frameworking"; development methodologies: inter-disciplinary platforms, evolving and adaptable approaches balancing structure and process concerns.

Also other areas of research need to be more specialized as to be applied to e-manufacturing. Knowledge management is one of such areas, particularly research work on ontologies, domain and aplication ontologies, ontologies and information systems. Even in the field of management information systems there is still work to do in inter-organisational information systems, distributed decision support systems, collaborative work support systems, negotiation support systems, specifically addressed to e-manufacturing.

5. REFERENCES

Barbedo, F., Soares, A., Sousa, J., Combining social structure and process analysis in collaborative networks. Proceedings of the IFIP Working Conference on Virtual Enterprises, PROVE 2003.

Bijker, W.E., T.P. Hughes and T.J. Pinch, The Social Construction of Technological Systems: New Directions in the Sociology and History of Technology, 1987. Cambridge, MA: MIT Press.

Castells, Manuel, The Rise of the Network Society, Volume 1 of The Information Age: Economy, Society and Culture, 1996. Oxford: Blackwell.

Cheshire Henbury, from its web site: http://www.cheshirehenbury.com, 2001.

Gittell, J., "Organizing Work to Support Relational Coordination," International Journal of Human Resource Management, 2000, 11(3):517-534. Harvard University Press.

Howard, P., Network ethnography and the hypermedia organization: new organizations, new media, new methods. New media & Society, 2002, 4, Vol. 4.

Jonkers, H. L.; Donkers, H. W.; Diederen, P. J. M. - The Knowledge Domain of Chain and Network Studies. KLICT. September 2001.

Lamb, R., Sawyer, S., and Kling, R. (2000) "A Social Informatics Perspective On Socio-Technical Networks," in H. Michael Chung, ed. Proceedings of Americas Conference on Information Systems, Long Beach, CA.

Lambert, Ad. M.; and Coper, M. C. – Issues in Supply Chain Management. Industrial Marketing Management. 2000. N° 29, pp. 45-56.

Latour, Bruno (1987). Science in Action: How to follow scientists and engineers through society, Cambridge, MA:Harvard University Press.

Magnusson, Johan; Nilsson, Andreas - To Facilitate or Intervene – A Study of Knowledge Management Practice in SME Networks. Journal of Knowledge Management Practice. February 2003.

Moller, K. K.; Halinen, A. – Business Relationships and Networks: Managerial Challenge of Network Era. Industrial Marketing Management.1999. N° 28, pp. 413-427.

Mowshowitz, A. On the theory of virtual organization, Systems Researchand Behavioral Science, 1997, 14(6), pp. 373-384.

Podolny, J. and K. Page, 'Network Forms of Organization', Annual Review of Sociology 24, 1998.

Trienekens, Jaques H., Views on Inter-enterprise Relationships. 2002, Department of Management Research. Wageningen University. The Netherlands.

Uzzi, B., 'The Sources and Consequences of Embeddedness for the Economic Performance of Organizations: The Network Effect', American Sociological Review 61: 674–98, 1996.

Walsham, G. and Sahay, S., "GIS For District-Level Administration In India: Problems And Opportunities", MIS Quarterly (23:1), 1999, pp. 39-66.

Wellman, B., J. Salaff, D. Dimitrova, L. Garton, M. Gulia, and C. Haythornthwaite, "Computer Networks as Social Networks: Virtual Community, Computer Supported Cooperative Work and Telework" Annual Review of Sociology, 1996, 22:213-38.

Zack, M., Researching Organizational Systems using Social Network Analysis, 33rd Hawaii Intl. Conf. on System Sciences, 2002, Vol. 7.

E-MANUFACTURING IN EUROPE: ENTERPRISE NETWORKING

Francisco J. Restivo
Universidade do Porto, Faculdade de Engenharia
IDIT – Instituto de Desenvolvimento e Inovação Tecnológica
fjr@fe.up.pt

In the emerging global economy, where the demand for mass customized products with very short life cycles put new requirements in terms of the ability of enterprises to respond to change, new working relationships are now as important as new manufacturing devices and new manufacturing control architectures.
The author presents a short overview of e-Manufacturing in Europe and the role of enterprise networking in achieving a highly competitive industrial environment.

Keywords: e-Manufacturing, Enterprise Networking

1. INTRODUCTION

In the emerging global economy, the demand for mass customized products with very short life cycles put new requirements in terms of the ability of enterprises to respond to change [1].

Enterprises always looked for more productivity, more quality, more flexibility and more agility, but used to focus in steady state operation.

Nowadays, steady state analysis is becoming less important, while dynamic response is becoming a key issue in the design of new manufacturing systems, with large impact at all levels, both technical and socio-economical.

The way manufacturing systems handle unexpected internal and/or external changes, learn changing patterns and change their behaviour to respond to new demands are basic parameters for the evaluation of enterprise performance.

New working relationships are now as important as new manufacturing devices and new manufacturing control architectures.

2. THE CURRENT SITUATION

Max-Serv, the Online Manufacturing Excellence Centre[1], states that European Process Industries competitiveness is only average when compared to World Class:

---

[1] http://www.max-serv.com

- Less than 4% achieve World Class performance in the majority of measurement parameters
- A first estimate of the potential for increased sales is €160 billion per year
- The bottom 25% of European companies receive only 50% of their supplies on time
- The bottom 25% of European companies require more than twice as many indirect employees as do the best.

But what is World Class Manufacturing?

Recently, one vendor[2] of manufacturing planning resources published a 20-point checklist, which included in first place:

Q1: Do you ship to your customers on time in full (OTIF), more than 99% of the time, against their latest schedule or delivery agreement?

Max-Serv has researched this issue and found the answers, some of them given in Table 1:

Table 1 – Measuring world class manufacturing

Parameter	European	World class
OTIF	93.1%	>99.1%
Complaints	2.2%	<0.001%
Accidents	$7.4/10^5$ h	$<0.05/10^5$ h
Finish goods cover	21.2 d	3 d
Raw material cover	28.3 d	5 d

Table 1 shows the size of the performance gap that has to be overcome, how far should European companies go to achieve World Class status.

3. STRATEGY FOR IMPROVEMENT

The answer to the challenge of reducing the performance gap relies in first place inside each company:

Although there is no prescription to improve performance, the following are critical steps to be taken

- Discover hidden capacities, at the human resources and the material resources level, before increasing resources number
- Reduce variable costs, which are always present n the final price, even for large production volumes
- Reduce maintenance and fixed costs, which affect largely final prices, specially for small production volumes
- Improve stock turn, with better forecast of customer demands and increased flexibility of production capabilities
- Increase added value per employee
- Improve safety, health, and environmental compliance.

[2] http://www.bpic.co.uk/wcm.htm

The emergence of the digital business is changing the situation, requiring new answers to be given. Now

- Competition is global, and customers want competitive products and services at ever-lower costs
- Companies are not able to answer to customer demands on a stand alone basis, needing to build up working, agile networks of partners
- Process automation and outsourcing are not sufficient, and partnerships require enterprise wide communications to be able to guarantee efficient cooperation between partners
- Companies compete on their ability to develop and use knowledge from the vast information flowing through networks.

So, companies, and specially SMEs, face new dilemmas, new problems they were not ready to cope with

- Multiple information systems, since SMEs are usually not in a position to impose their own solutions, but they have to be ready to work with the information systems of their partners and/or customers,
- Know-how not available completely inside, as products become more complex and responsibilities are pushed down the supply chain
- Lack of ICT solutions targeted to SMEs
  - For distributed design
  - For distributed planning
  - To support the networked enterprise.

Companies need to understand their business and to manage their decision making process. Obvious requirements are

- An enterprise framework where business can be managed, where decisions can be made and implemented
- Business driven, rather than IT driven ICT solutions
- Solutions allowing for changes in business rules without costly rework
  - Of applications
  - Of infrastructures.

The last requirement is becoming absolutely critical. World is dynamic, there is no steady state, and decisions cannot be based in mean values but in signals.

Company's objectives have to be translated into optimisation problems over their full life cycle.

4. INTEGRATED VIEW

In a healthy economical environment, enterprises should be born, live and die naturally.

If society expects companies to live forever, to keep jobs forever, as if they had indefinite life, then companies fulfil a social role they have not been created for.

The requirement for indefinite life has to be dropped, allowing companies to take full control of their life cycle.

However, such a simple statement hides a completely new paradigm, where the role of enterprises and of the state is redefined, and the commitment of enterprises to indefinite duration working posts is released without social drama.

4.1. The enterprise

An enterprise uses materials, energy and information to deliver products and/or services. The transformation requires resources (material, human), generates waste, and is subject to restrictions (quality, environment, health and safety).

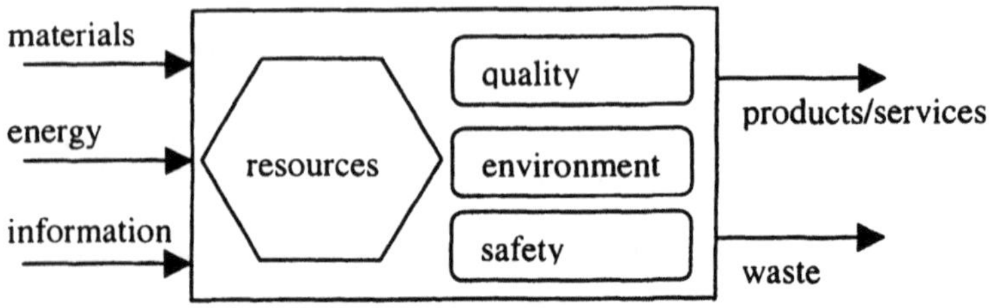

Figure 1 – An enterprise view

The enterprise is an intelligent system, which learns and develops knowledge, responds to change and closes when cannot meet the objectives it has been created for.

Common requirements are flexibility – capability to produce a range of products, agility – capability to change from one product to another, and ecological efficiency.

4.2. The networked enterprise

The networked enterprise is an interconnection of enterprises with a goal, but where the networked enterprises retain their own decision capabilities. It can take a number of forms – supply chain, extended enterprise, e-alliance – to name but a few.

SMEs are facing new demands from large companies that they cannot satisfy alone. Networks are often the best way for existing SMEs to respond to the changing demands of the environment.

Setting up alliances/cooperation networks between SMEs should be part of a strategy for business development, such as

- Bottom-up: marketing synergy for reaching new markets
- Top-down: subcontractors partnership towards main contractor
- Horizontal: resources/means sharing for higher efficiency.

Quite frequently, large companies don't want to work with a large supplier base, pushing their suppliers to set up networks. In other situations, networks are the answer when large companies require that their suppliers have a strong customers base to assure sustainable relationships.

In general, large contractors require from their suppliers

- Global solutions: providing complete subsets of products/equipments
- Capacity of innovation: developing a link to co -contracting

- Global performance: guarantying objectives, achievement, reliability, and sustainability.

On the other hand they expect from their sub-contractors

- Operation in an industrial network
- Improved competitiveness for the main-contractor
- Enriched / full service package: manufacturing, purchasing, design, etc.
- Integration of new specific added value: skilled competence, complex project management, etc.

4.3 New needs

More and more SMEs are assuming a worldwide view from their local position onwards, and are getting involved with ICT tools and methods. But they still need

- Awareness before any kind of co-operation starts
- Actions organised in a local environment
- Basic training customised to their individual situation (e. g., project management).

And once they are engaged in some form of cooperation, they need assistance in

- Participating with R&TD Centres in R&D projects
- Implementing R&D project results (ICT methods and tools)
- Facilitating changes in this new networked environment.

European IST project e-Power – Powering e-Europe's Regional Economy[3] aims to establish a network of RTCs – Regional Technology Centres - for RTCs. The e-Power network will optimise the regional understanding of the RTCs allowing them to communicate to their local SMEs the business benefits of adopting new ICT technologies and to respond to the new needs that are arising. Particularly attractive is the integration of regional RTC initiatives and experiences into this information flow.

4.4 The new paradigm

Enterprises of the future face new challenges that cannot be met with the answers of the past.

The new paradigm may be a large resources marketplace, which sustains the new, distributed intelligent networked enterprises

- With innovative business strategies
- Business driven
- Competing on knowledge, agility and cost.

The performance management system – performance planning, performance measurement and performance improvement – is a key part of these new enterprises [2,3].

A measurement methodology has therefore to be developed according to some rules, as for example [4]

[3] http://www.e-power.info

- Performance measures have to be chosen in harmony with company goals.
- Performance measures have to enable comparison of organizations that work in the same branch.
- Use of each of performance measures has to be clear.
- Collection of data and calculation methods has to be clearly defined.
- Performance measures should be chosen based on discussions with persons, who are connected with these measures (customers, employees, managers).
- Objective measures should take precedence over subjective measures.

Performance measures for enterprise agility or flexibility have however still to be defined or agreed.

5. CONCLUSIONS

The author intended to show that new economical paradigms where enterprises are able to respond in real time to environment changes would emerge, but that such a major change will require that many long accepted issues be revisited.

The ability to build up immediately high performing enterprise networks, supported by a performance management system is a critical success factor.

6. REFERENCES

1. Leitão, P. and Restivo, F.: Identification of ADACOR Holons for Manufacturing Control. To appear in Proceedings of 7th IFAC Workshop on Intelligent Manufacturing Systems, Budapest, Hungary, 6-8 April (2003).
2. Seifer, M.: Automated Performance Measurement (APM) in Extended Enterprises. In Proceedings of the 2nd IFIP International Workshop on Performance Measurement: Performance Measurement for Increased Competitiveness, Hanover, Germany, 6-7 June (2002).
3. Chrobot, J.: Performance measurement of manufacturing systems – requirements for a tool. In Proceedings of the 2nd IFIP International Workshop on Performance Measurement: Performance Measurement for Increased Competitiveness, Hanover, Germany, 6-7 June (2002).
4. Globerson, S.: Issues in developing a performance criteria system for an organisation, International Journal of Production Research, Vol. 23 No. 4, page 639-46 (1985).

KNOWLEDGE, INNOVATION AND INVESTMENT

José Manuel Mendonça
Faculty of Engineering – University of Porto
Ilídio Pinho Foundation

Motivation

This paper is meant to share some views on the relevance of science and R&D to business development and to sustainable economic growth. While attempting to do this, it is also meant to present some insight to an encompassing and complex reality with a multitude of interacting actors with different values, motivations, time-frames, cultures and even language: scientists, technologists, entrepreneurs, brokers, investors, politicians, etc.
In this paper, knowledge is first discussed as the competitive advantage of our time. Then, the "knowledge-to-competitiveness" value chain is presented and discussed. Being innovation a key concept, the various types of innovation are discussed and the use of some major innovation management tools is briefly introduced. The business perspective is then put forward, discussing the complex mechanics of knowledge management scoring, due-diligence and rationale supporting investment decisions in novel science-based companies.

Keywords: Innovation Management, Knowledge Management Scoring, Due diligence

1. KNOWLEDGE: WHAT IS DIFFERENT IN 2002?

Some say that our era is the one of the knowledge-based economy. Others refer to it as becoming more and more the era of a digital economy. Others still name the novel businesses based on the digital products and services as the "new economy", as opposing to the traditional or "old economy".

It is fair to acknowledge that information and knowledge have been winning factors since ever in warfare, in politics and in business. And exactly the same applies to a more or less tangible result of scientific knowledge, we can designate by technology, this being specially true since the first industrial revolution. The relevance of information, knowledge and technology is thus certainly not a novelty of the present times.

But it should also be acknowledged that the business globalization brought in by the fall of nearly all trade barriers, together with the very fast penetration of information and telecommunication technologies and the global computer networking, resulted in:

- an increased importance of information and knowledge, but now with requirements of instant access to real-time updated, state-of-the-art certified information;
- an acceleration in the process of knowledge acquisition, processing (transformation) and distribution (diffusion) that, in turn, fosters faster changes in all aspects of business and in nearly all activities of society;
- a specific aspect is the acceleration of the advances in science and the consequent acceleration of technological changes
- the emergence of a critical competence: that of being able to use both tacit and encoded knowledge for improved decision making at all levels as to improve company effectiveness and competitiveness.

This acceleration has been made possible thanks to the huge amount of resources that have been put together both by public and private funding in all developed countries.

Public R&D expenditure has been driven by strategic sectors in any country, such as health, aerospace and defense, while supporting fundamental research as well. Private investment in science and technology has been essentially market driven. Pharmaceutical, chemical, agro-food, semiconductor, consumer electronics, software and computing and automotive have been leading sectors in private R&D investments.

These high-tech global industries were the first ones to perceive and take advantage of the globalization: in fact, global outsourcing and e-business have been used for decades by automotive and pharmaceutical companies. These companies also support their science-based, technology intensive, manufacturing through the economy of scale of a global operation. However, this R&D effort was pushed so far in the last 10/15 years that even the largest corporations in those sectors were forced to joint-ventures, merges and strategic alliances with their former direct competitors.

Knowledge, not only embedded in technology – materials, manufacturing equipment, processes – as well as in manufacturing organization and product or business models – plant layout, logistics, CAD/CAE, outsourcing – was then transferred to the first-line suppliers of these high-tech global companies and later spread into the whole of the value chain.

Medium and low-tech manufacturing sectors - such as textile and garments, shoes or moulds - became global businesses of increasing competitiveness, as well, and had to incorporate rapidly both advanced technology and novel organization models.

At the turn of the century one can be sure of:

- the trend of science-based manufacturing to widespread rapidly to nearly all sectors;
- the absolute need to support investments in science and technology through a large business base, given the huge size of the investments needed.

Some claim that one of the causes of the US superiority over the ex-URSS was actually closely linked to the very essence of the last statement (see Figure 1).

In the former Soviet Union, R&D was financed solely by the state and was driven almost exclusively by the military and aerospace sectors which were state owned and controlled. The output of the work undertaken by the universities and state laboratories was often classified and kept confidential within these ivory towers with no relation whatsoever with the other sectors of economy and society.

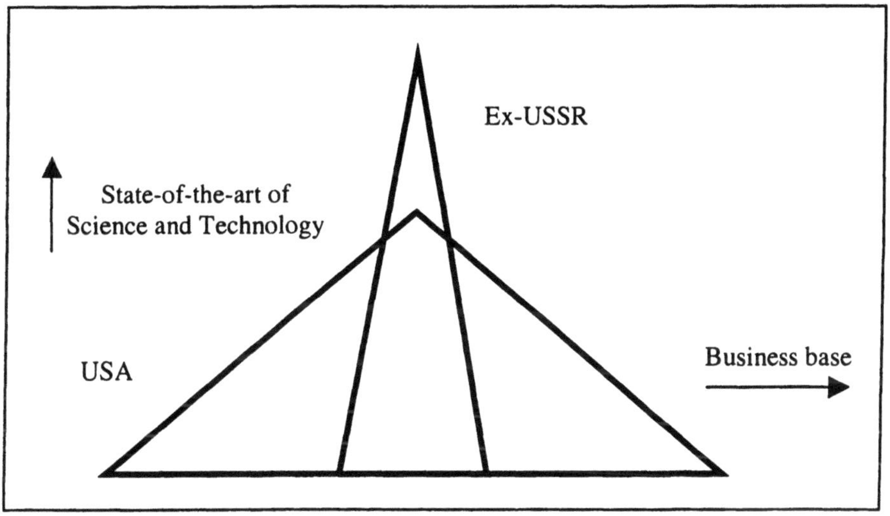

Figure 1 - The ivory towers of knowledge versus the knowledge-intensive businesses

In the US, the same technology pull through military and aerospace applications has occurred as well, but it was soon shared and disseminated through the participation of many high-tech companies selling different products in other markets as well. The science and technology base acquired was therefore spread out and made useful to other lines of business, increasing their competitiveness and leading to profits that could also foster new R&D developments. Through knowledge diffusion, private investors in R&D could thus take advantage of public investment to foster competitiveness in many other areas of business. Such public-private complementariness has been a winning factor.

Instead, in the ex-URSS the state alone had to support the enormous financial effort needed to be at the state-of-the-art of science and technology in many different areas and that became quite an impossible task without the market valorization of the R&D results produced. In fact, in both cases, through investment in R&D, "money was turned into knowledge" (ivory towers R&D) but only in the former case "knowledge was turned into money" (business oriented innovation).

2. THE KNOWLEDGE-TECHNOLOGY-INNOVATION-COMPETITIVENESS LIFE CYCLE

A very brief remark on the issue of knowledge sharing, learning processes and the "knowledge-technology-innovation-competitiveness" life cycle will now be presented.

The issue of tacit versus encoded knowledge, which has already been referred to above, is now revisited (see Figure 2). In fact, explicit knowledge encoded into patents, engineering documentation, software, etc., plays a critical role in all technology-based businesses. This science-originated knowledge is fundamental if a company is to achieve product differentiation and it is quite simply and cheaply distributed through CD-ROM's or remote data base file access. As it is easily transactioned it may require adequate protection. The encoded science-based knowledge, closely linked to the academia, is produced in universities, research institutes or company research laboratories.

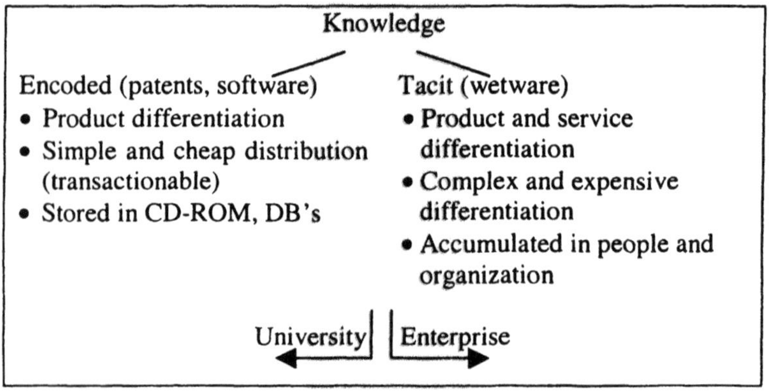

Figure 2 - Knowledge sharing and distribution

In a company, this type of knowledge needs however to be complemented by the tacit knowledge (often called wetware) that people and organization have, both on the actual application of the encoded knowledge and on other detailed implementation issues. This is an experience-based knowledge and its contribution to the differentiation of products, processes and services is of utmost importance as well. This type of knowledge is however of a more complex and difficult distribution: its transaction usually involves transferring people. The tacit knowledge is usually generated within the company, but sometimes it is often captured by consultants and scientists working closely with the company.

The encoded and the tacit knowledge also differ in the learning processes that they require (see Figure 3). The first one requires the formal learning processes used in higher education or the scientific methods of experimental research. This is obviously the world of researchers and academia. Tacit knowledge requires informal learning practices based on day-to-day on-the-job experience and informal interaction. This is naturally the enterprise environment.

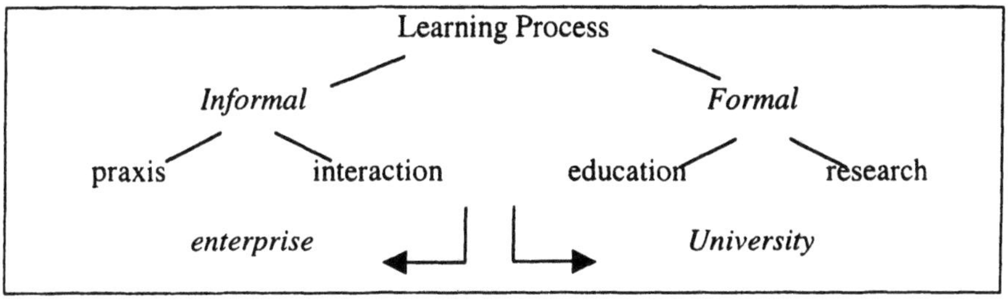

Figure 3 - Knowledge creation

The life cycle, or value chain, in which knowledge is the source or primary raw material, encompasses both the research environment and the enterprise and business world. Although it is nowadays risky to draw borders separating which activities should be undertaken in the university and which do belong to the enterprise world, a simple model may help us to understand when, where and how is the knowledge transfer taking place (see Figure 4).

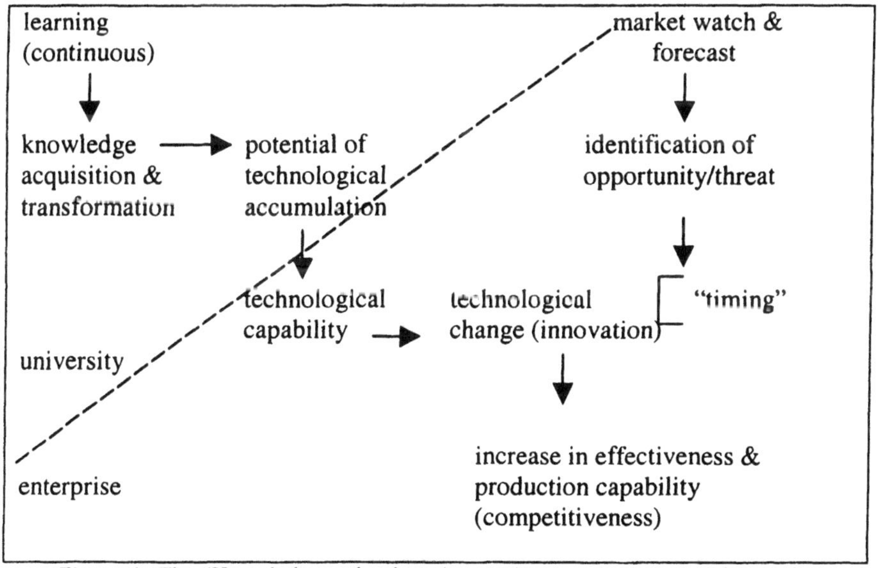

Figure 4 - The "Knowledge-technology-innovation-competitiveness" life cycle

The continuous development of new knowledge/science from acquiring and transforming existing knowledge/science generates a potential of technological knowledge. Such technological knowledge may source a technological capability in the specific area of expertise, which can (eventually) be made useful. This technological capability may then be used by companies to trigger a technological change geared to the improvement of existing processes, products or services. However, this will only be done when, upon the close observation of the market and the competition, an opportunity or a threat is identified requiring change for improvement, that is requiring innovation. Such technological innovation will always be directed at improving effectiveness, for example through differentiation brought into by new design and/or manufacturing capabilities, thereby fostering competitiveness. So, for the technological capability to be fully used as the trigger to technological innovation, a timely market opportunity is needed.

3. INNOVATION: A RECIPE OR AN ATTITUDE?

Innovation is one of the strongest buzzwords of the last decade. Scientists, scholars, entrepreneurs, politicians and journalists use the word, certainly in many different ways, with a common though vague idea of improvement through a new way of doing things.

From the 50ies into the 90ies, companies found increasing demanding markets and fierce-full competition (see Figure 5). At first, to offer products (functionality) was just enough to stay in business. In the 70/80ies, cost-quality-time requirements became mandatory. Nowadays, meeting those requirements is no longer sufficient to ensure company's competitiveness. Innovation, as a differentiation factor to products and services, became a major issue. Actually, mass production is giving way to the so-called mass customization (the consumers want the same product but each one of them with own single features and the vendors have to meet that demand; Dell computers, Motorola pagers and BMW cars are examples in different high-tech industries).

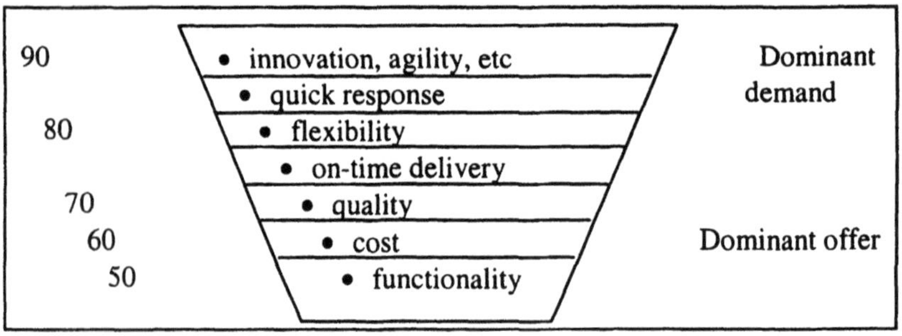

Figure 5 - Increasing demands foster competitive pressure

The semantics of the word innovation may suggest invention, change or improvement. Perhaps the most simple but powerful definition is "innovation is profitable change". Profit here should be obviously understood *latus sensus*, not just in the economic sense but also in the more general sense of improvement (economic, environmental, social, etc.).

Depending on its main focus and position in the value chain, a company may be innovative in products, processes or services. Usually these types of innovation are related; for example, product innovation always calls for (manufacturing) process innovation and requires often service innovation; and service innovation quite often demands (logistics) process innovation.

Depending on the main origin of innovation, it may be seen as essentially technological or predominantly organizational. In the first case, changes in technology are determinant, whereas in the later case it is the organization of the resources that prevails. Once again, there is a close relation: whenever adopting novel technologies it is often mandatory to re-design or to re-optimize the organization both in manufacturing and in services.

Finally, depending on the rate of change which is being imposed, innovation can be said to be incremental or radical. Incremental innovation is the low risk approach of adopting step-by-step changes of small individual impact, thereby ensuring a steady though firm improvement trend. Radical innovation is linked to important and sudden changes to the core of the company's activities. It is often a high risk but necessarily a high reward strategy.

In the life of an enterprise all these types of innovation may occur and they all relate to each other. But the enterprise has to ensure stability periods of operation as well, as to maximize the efficiency of their manufacturing and business processes. The Japanese three colors KAIZEN model (see Figure 6) shows how this balance between change and stability can be achieved and managed. The left (dark grey) triangle means that roughly 50 % of the time of the company is in the so-called business-as-usual mode. The small (soft grey) triangle on the upper right represents the time and effort that the company should devote to radical, usually both technology- and organization-based, innovation. And the (white) area in between represents the everyday effort of incremental innovation: the continuous improvement (in Japanese, KAI ZEN).

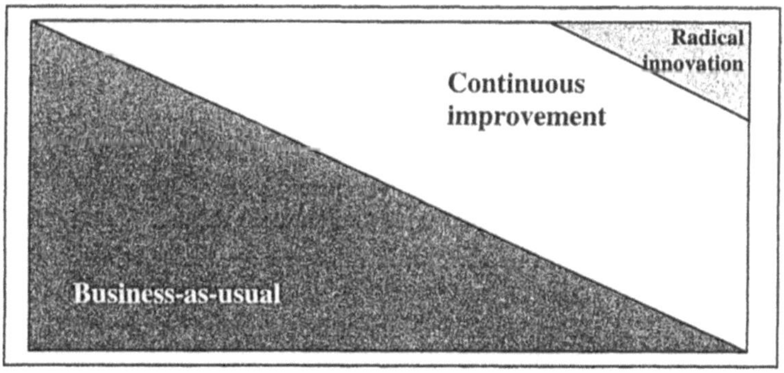

Figure 6 - KAIZEN three-colour model

This discussion has been focused on manufacturing companies and on how they undertake innovation in their products, processes and services with the aim of reinforcing competitiveness. But the essentials apply to nearly all service sectors as well: health, banking and insurance companies, retail commerce, large-scale distribution, teaching, public administration, etc. Most of these sectors still have a lot to learn from the world-class manufacturing companies and businesses, for example in planning, logistics, resource management or quality assurance. But they can also learn from the SME's that have to compete globally with a much more limited access to resources, in areas such as procurement, outsourcing, short lead times or client orientation.

If we were to push this argument to the limit, we would emphasize that although innovation seen as "turning knowledge to money" is essentially a key factor in a market-driven business strategy, the attitude of being innovative is a distinguishing feature of regions, companies and even individuals. And the motivation and capability to improve how things are done are obviously very much linked to information and knowledge and very much dependent on these assets.

4. INNOVATION MANAGEMENT TOOLS

Innovation management is a rather complex issue involving many different aspects, including innovation protection, dissemination and transfer (most of the times people refer to technology transfer instead, which is more tangible but less encompassing).

Another source of complexity arises from the fact that innovation is usually highly multi-disciplinary; product innovation, for example, often involves expertise in design, materials, mechanical, electrical and electronic engineering, as well as in marketing, product costing, etc. Furthermore, product innovation is a process requiring the use of specific management know-how, such as simultaneous or concurrent engineering.

Due to its encompassing nature, company innovation capabilities are also strongly dependent on the quality and level of education and skills of the human resources, from shop-floor operators to engineers and top-managers.

Finally, innovation is stimulated by some types of leadership as well as by the company organizational environment. It can actually be said that innovation is a matter of talent, which is able to position the (already) efficient companies in a superior level of effectiveness.

Different innovation management tools are used in the various phases of the life cycle: R&D result, laboratory prototype, industrial prototype, production pilot or routine production. The value of innovation grows as one evolves from the early stages to the end of the life cycle. This stems from the fact that the risk decreases because the use of the technology, the inherent organizational impact and the associated economic value in terms of market potential become much better known.

A simple value-chain model is portrayed in Figure 7, where different innovation management tools and their inter-relations are shown in three so-called domains of influence: innovation offer, innovation demand and innovation mediation. On the innovation demand side, this being certainly the case of any industrial SME, innovation audits, technology audits and benchmarking studies are diagnosis tools that help looking at problem identification. On the innovation offer side, the domain of the science and technology producers, important tools are technology rating, technology watch and forecast and intellectual property rights (IPR) protection. Finally, the mediation domain uses tools aiming at facilitating the offer-demand relationships, such as e-marketplaces for technology, brokerage events & face-to-face follow-up meetings, technology clinics, best practice identification, technology licensing and spin-off launching.

The use of these tools will hopefully lead to innovation pilot projects of different types, which are naturally of mutual benefit for companies and for research centers. From the demand side, the benefits in improved competitiveness have already been discussed. From the university and research center side, the improved motivation brought-in by the valorization of the R&D work and the recognition of its social/market relevance, as well as the extra funding from the contract research, licensing or royalties are most important benefits.

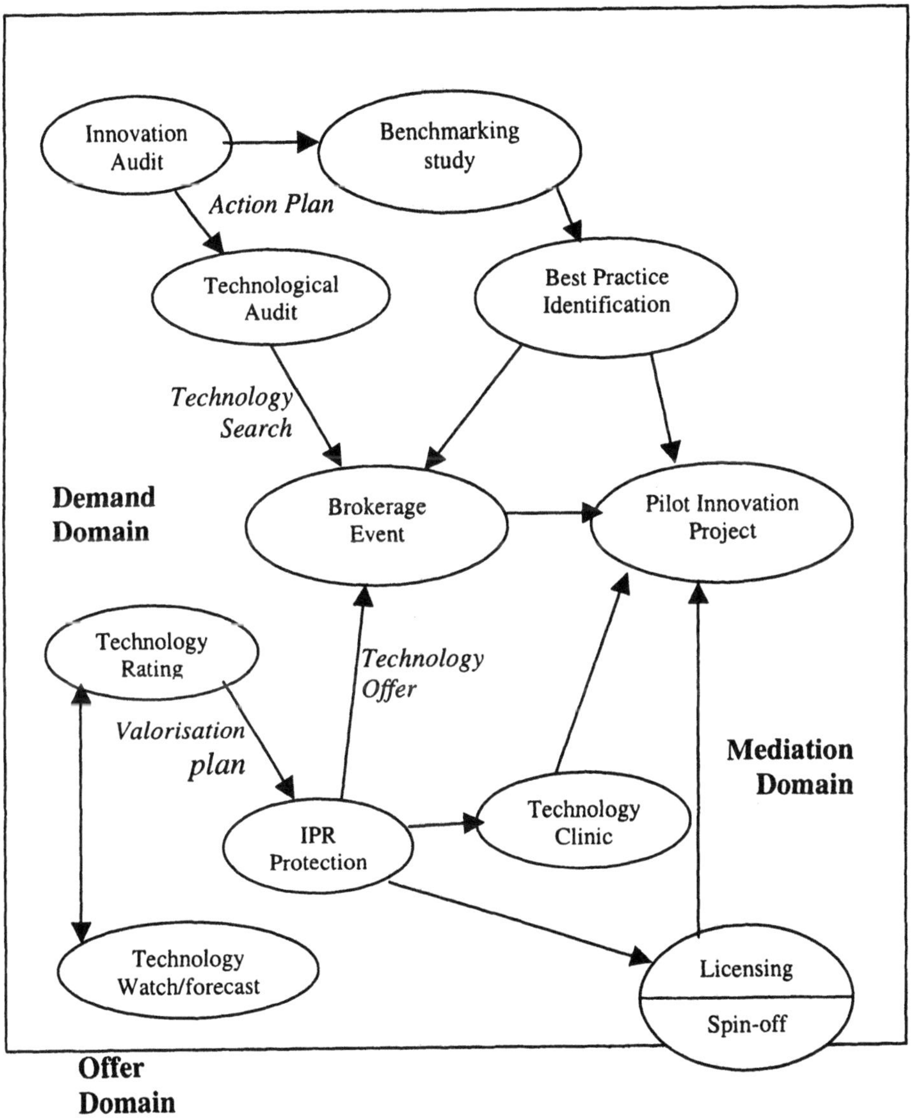

Figure 7 - Innovation management tools in the value-chain model

5. KNOWLEDGE MANAGEMENT SCORING

Companies – certainly the management but often the shareholders as well – have to undertake the evaluation of their investments in innovation, either internally or with external help. A company with a very good control of their core processes, a fair knowledge of the market and the competition and some insight in the best practices in the critical areas, is quite able to perform a good assessment of the costs, return on the investment, risk factors, etc., which are inherent to a particular innovation project.

However, for an external entity – a financing bank or a potential investor – to undertake an assessment of an innovation project or of the value of an innovative company with considerable intangible assets, a substantially larger effort is required.

The need for scoring the intangible as well as the tangible assets of a company is now more important than ever. The value of the IPR's and licenses can be estimated through present and future sales and placed in the balance sheet. But the value of the tacit knowledge – part of which is the market value of the human resources – is much harder to evaluate, if at all possible.

Knowledge management scoring is intended to allow the balance sheet to include the value of encoded knowledge, the value of tacit knowledge and other intangible assets. The market does this everyday, but formal tools to support such an assessment are still in the research domain.

6. DUE DILIGENCE IN THE KNOWLEDGE ECONOMY

With scarce decision-making tools available, the investor uses a mix of objective and subjective criteria to support the investment decision. Bank cashier type investors look at personnel and equipment costs, sales, margins and return on investment, often only as far as they can see, that is in the short term. Real venture partners do more than that: they share the long-term vision and the business strategy of the entrepreneur and sometimes they even help him to build that business strategy. Venture partners often have strategic interests in the area and they are often able to sell services and to participate in the management. But their life is not easy as well.

Market assessment in emerging areas, such as genetic-based diagnosis, advanced UMTS services or nano-manufacturing is a rather specialized (and risky) job. The required technology assessment, as well as technology watch and foresight, is an in-depth activity calling for specific expertise.

This is why many investment banks and risk capital companies may be unable to undertake due diligence in most of the knowledge-intensive businesses. Outsourcing the necessary world-class competencies while ensuring the needed independence may not be easy as well, at least in many countries.

Conventional investors are experienced in well-established conventional businesses that they can understand and they accept running risks in the areas where they move easily. They more seldom invest in less tangible businesses – wind farms, molecule value-adding, novel construction materials, etc. – that are hardly understood by them. It is actually difficult to estimate the market size, competition, potential market share and their evolution, since only small niche markets are starting to emerge in some developed countries.

In fact, extensive and in-depth market studies are a too expensive exercise for most of the new-technology-based firms and businesses. And if these are over-simplified there is a huge risk that they will tend to generate over-optimistic expectations, often the pure guessing exercises that many business plans include.

Besides problems such as technology lack of maturity or slow market take-up, regulatory or certification issues may convey further uncertainty to the novel technology-based businesses. This is the case with the grid connection licenses to be granted to wind energy parks; the legal/ethical barriers relating to the genetics-based development of human tissue or parts of organs; or the safety legislation related to the adoption of novel composite materials in the construction industry.

All that adds up to a level of difficulty that only large multinational corporations can manage thoroughly. For almost everyone else, the risks of a "wrong" decision grow rapidly. The very fast up rise of the so-called internet stock market "bubble", with nearly everybody investing in emerging companies of unknown shareholders launching misunderstood businesses, was a dramatically good example of the risks of being wealthy though uninformed.

7. THE DIFFERENT PARADIGM OF THE FINANCIAL MARKET INVESTORS

While company or business assessment is a knowledge-intensive task, where many aspects matter, even the psychological profile of the entrepreneur, investing in the financial markets is conducted through a totally different paradigm.

When investing in the financial markets, most of the times the on-line decisions just follow trends: there is no time to collect knowledge, there is no time to analyze, only to take action. When telecom companies go down, just sell what you have (including the shares in sound, profitable companies). If biotech is going up, just buy everything you can and fast (even the shares of lousy companies that you really do not know about).

Investment is therefore guided by trend analysis of on-line information on company past behavior (dividend distribution, control attempts, etc.) and not by company and/or business specific in-depth knowledge. While trading requires a huge effort in information management it includes negligible company specific knowledge management. And this is taken to the limit when investment decisions are made at the level of edge funds, structured products, mixed portfolios, etc. There is a huge distance between reality and those making decisions on that reality. And such a distance has increased, because within the global economy much more investment decisions have to be made in much less time with (necessarily) incomplete information. The increased rate of change and the need for instantaneous decisions decreases human perception and calls for sophisticated computer models to support decision-making. Two consequences arise: first, since all the traders use the same models and they all use the same real-time information to support all their real-time decisions, the added up result is often that their decisions cause the trend up and the trend down leading to irrational overshoots. Secondly, the human is put out of the decision chain and the computer models simple assume that accounting cosmetics is not done because it is just not legal ... and therefore the computer models will be reasoning over wrong axioms.

Finally, investment through funds built on mixed portfolio increases security by spreading risks, but further increases the distance between the investor and the companies he is investing in. The investor just knows the name of the fund, its track record and the fund manager past performance.

8. SMALL SIZED PROJECTS OF YOUNG ENTREPRENEURS: WHO CARES?

Science-based innovation may successfully reach the end of its value-chain in quite different environments. Within the borders of a large corporation, it is very unlikely that an innovation project in the company's core business will not be able to find internal financing. In a high-tech well-established SME, external funding may have to be sourced but innovation projects are not likely to stop as well. But what about the start-up company with a novel good idea, knowledgeable people, technology and a fairly good business plan but with scarce financial resources ?

Within this scenario who will want to invest in small new technology-based firms (NTBF's) launching knowledge-intensive high-risk projects supported by science-based innovation? Reasonable sized investors want reasonably sized business opportunities with a proportionally low cost of the due-diligence and of the investment follow-up. And investors want to pre-negotiate conditions to leave the company and make money through selling options, IPO's, etc.

This is not what happens with NTBF's and with knowledge intensive spin-offs which are high risk and often do not assure high volume business. These have to attract venture capitalists with guts feeling and intuition as well as with discretionary power.

In Portugal, a breed of small NTBF's have made their way: niche markets, high-tech science-based businesses and international partnerships with clients are the key success factors. Certainly not all made it: the profile and management capabilities of the entrepreneur are of critical importance.

For building a business plan and launching a new technology-based business, different approaches are possible: they are often referred to as the business-school, the engineering or the high-tech approaches. But, in fact, any of these need to mix with what one could designate by "the guy that went nuts and wants to try something crazy" approach to launching a new high-tech business.

Experience shows, though, two risks for those companies once they start growing. One is related to the research-to-business transition: they have to stop investing in the next generation product and just sell the one they have now! The other is the growing strategy and partnerships: grow steadily using almost internal results or through discontinuities with bank loans, new financial partners or IPO's ?

9. CONCLUDING REMARKS

A few challenges for investing in knowledge are now left, leaving us with more doubts that those we have at start. First, can we afford being (only) non-conventional investors? Second, the due diligence in new technology-based businesses can be made (sometimes) locally (parochial) or should it always be global? Finally, does it depend on where you are doing it (country, continent)?

These questions can be raised in advanced country or regions and the answers will perhaps not differ so much from those obtained in Portugal.

JOINT VENTURES FOR INNOVATION & COMPETITIVENESS

José Carlos Caldeira
INESC Porto
jcc@inescporto.pt

Innovation is known as a very complex and multidisciplinary activity, calling for a wide range of competences. Although large companies can easily have access to resources (people, infrastructures, capital, etc.) needed to promote and manage Innovation, it is very difficult for SME's to cope with those requirements.
Strategic Cooperation Networks (SCN) are emerging as a valuable organizational tool that can create considerable competitive advantage for small firms, by establishing stable cooperation framework and by providing mutual benefits and results beyond those that a single organization or sector could attain by themselves
This paper presents some results of a recent research conducted by INESC Porto, aiming at the characterization and analysis of the Innovation process and the identification of critical success factors in terms of consortium structure and management.

Keywords: Innovation Management, Technology Transfer, Project Management, Corporation Network

1. INTRODUCTION

The success of Innovation - **"the act of putting into practice a recently invented process, product or idea, with a commercial exploitation purpose" (Schumpeter)** - depends on the success of individual Innovation projects. But an Innovation project is successful only if it leads to novel products, processes or services whose market exploitation has a positive impact. An improved assembly process, a novel line of products or the launching of a spin-off company usually results from successful Innovation projects.

Innovation is known as a very complex and multidisciplinary activity, calling for a wide range of competences. Although large companies can easily have access to resources (people, infrastructures, capital, etc.) needed to promote and manage Innovation, it is very difficult for SME's to cope with those requirements.

Strategic cooperation networks (SCN) are emerging as a valuable organizational tool that can create considerable competitive advantage for small firms, by establishing stable cooperation framework and by providing mutual benefits and results beyond those that a single organization or sector could attain by themselves. However, recent studies estimate that up to 60% of the alliances fail to meet their initial objectives [1], in a clear indication that there is still a lot of work to be done regarding the creation and management of these networks.

INESC Porto, together with the Science and Higher Education Observatory (OCES) of the Ministry of Science and Higher Education (MCES), has set up a research project to analyze and characterize the whole Innovation process and the relations between the different players. The most relevant functions/roles for the success of individual Innovation projects, the optimal consortium structure to maximize their success and the main critical success factors will emerge from this analysis and characterization.

2. OBJECTIVES

The main objective of our research is to focus on individual Innovation projects and to derive:

- The most relevant functions and players.
- The impact of consortia definition and function allocation on project success.

In order to support the analysis, correlations between the results of Innovation projects and the structure of their consortium were established. Moreover, the relationships between the entities that participate in an Innovation project and the functions played by each one of them were described. From here, the optimal consortium structure in order to maximize the success of specific Innovation projects was derived.

3. METHODOLOGY

The methodology used to develop this work has three phases:

- **Definition phase**: During this phase, the main functions, players and outputs/results of an innovation project were defined, along with a set of indicators and variables to characterize and measure them.
- **Data collection phase**: A survey was developed, based on the definition phase outputs, and the information was collected during direct interviews with the innovation project coordinators. A sample of 50 new product development projects was considered.
- **Data processing and analysis**: During this phase, the data collected was processed, and specific composite indicators were built, to measure and characterize the innovation projects, as well as to define correlations between project results, and different project organization and management examples.

4. INNOVATION PROCESS CHARACTERIZATION

The Innovation process characterization was based in three main vectors: roles/functions, actors, and results.

4.1 Functions/roles

Figure 1 represents our definition of the main Innovation process functions or roles (considering the objectives of this project), encompassing the whole life cycle:

- **Basic research**: experimental or theoretical work undertaken primarily to acquire new knowledge, without any particular application or use in view.
- **Applied research**: also original research undertaken in order to acquire new knowledge, but directed primarily towards a specific application or objective. It develops ideas into operational forms (ex. prototypes), often protected (patents, IPR, etc.).
- **Development**: includes the tasks of technology validation, product definition, specification and development.
- **Marketing and sales**
- **Maintenance and support**
- **Dissemination**: innovative products often need specific awareness and dissemination actions to promote their use.
- **Consultancy and training**: innovative products need specialized support.
- **Funding**
- **Coordination:** global project coordination

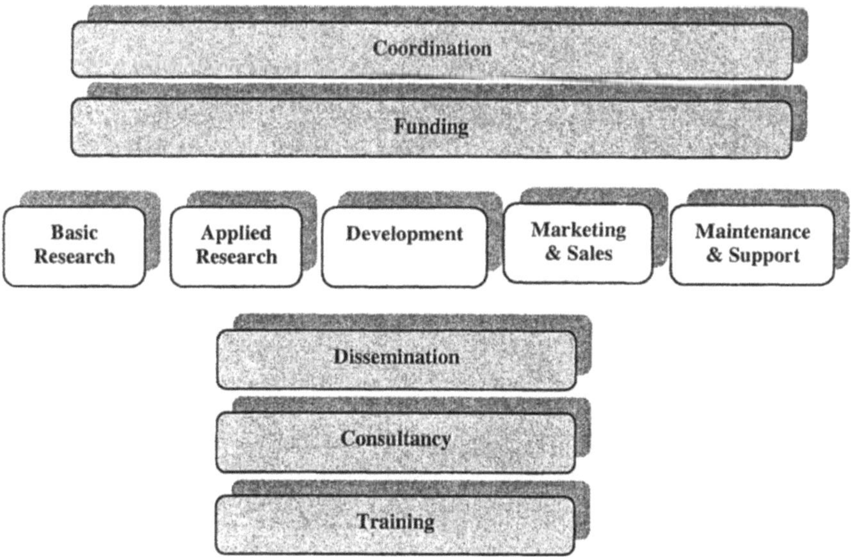

Figure 1 - Main functions or roles of the Innovation Process

4.2 Actors

In a similar way, Figure 2 organizes the main actors that can be involved in an Innovation process:

- **Universities**
- **R&D Institutes**
- **Technology Centers**: either sectorial or thematic.
- **Technology Brokers**: companies or institutions mainly involved in technology intermediation.
- **Products and Systems Providers**: machine manufacturers, software houses, system integrators, etc.
- **Service Companies**: consultancy, training and engineering companies, etc.
- **Sectoral Associations**
- **Users**: end users of project results
- **Funding Organizations**: managers and promoters of R&D and Innovation funding programs, business angels, venture capitalists or other investors.

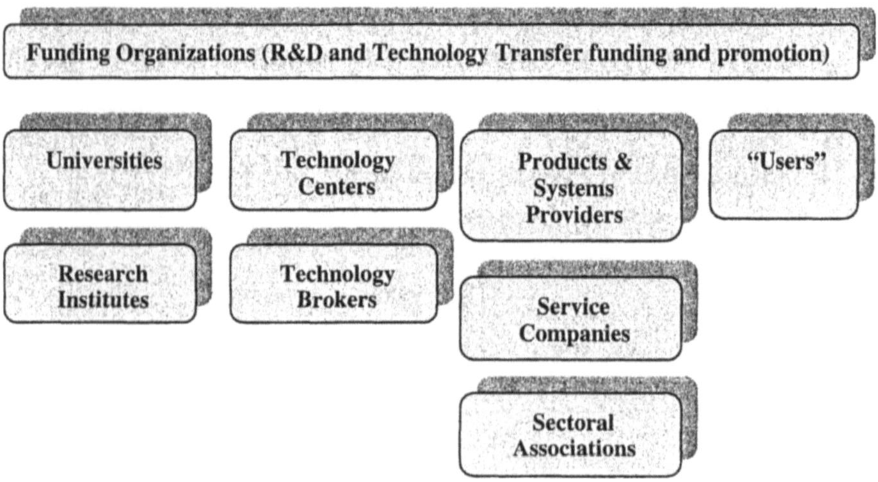

Figure 2 - Main Innovation process players

4.3 Results

The results of Innovation projects that concern our research are mainly novel processes or "products" developed and sold / installed. Other results, such as patents and publications obtained, M.Sc. and Ph.D. thesis and spin-off companies created were also analyzed.

Main Innovation process results

The indicators defined and the variables used to measure Innovation process results, refer mainly to:

- **Number and type of novel "processes" or "products" obtained per project.**
- **Number of units sold/installed per product/process and per project.**
- **Type of Innovation**: incremental or radical Innovation.
- **Sectoral impact**: whether results were transferred to just one or to a limited number of companies (small impact), to a large number of companies in the same industry sector (causing a significant sectoral impact) or to more than one sector (multi-sectoral impact).
- **Company or University spin-offs**: new companies resulting from the project.
- **The impact of project results for end-users and technology "vendors"** (companies responsible for project results exploitation), namely in the following areas:
  - Economic and financial ratios (turnover, ROI, productivity, etc.).
  - Market share.
  - Technology level.
  - Human resources qualification level.
  - Company image.

Other Results

Indicators and variables concerning other project results, regarding mainly the scientists involved, were also considered, namely:

- Number of patents obtained.
- Number of publications.
- Number of Master and PhD thesis.

5. ANALYSIS OF PRELIMINARY RESULTS

Although we have so far only collected information on 18 projects out of the total of 50 (mostly from the shoe and metalworking sectors), it is possible to extract some preliminary results that we consider relevant though they are qualitative in nature.

5.1 Function allocation scenarios

Traditionally, the "ideal" scenario for function allocation is the one represented in Figure 3.

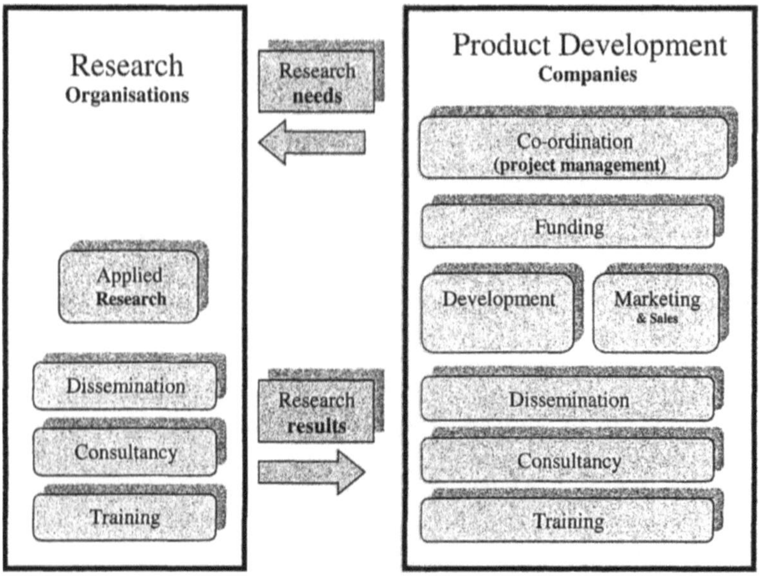

Figure 3 – Traditional function allocation scenario

Product development companies should be responsible for the execution of a set of functions, going from market research till marketing, sales and support, including also project and R&D management and, eventually some research tasks. This allows them to have control over project execution, to manage the interface with external R&D organizations (if needed) and to ensure an effective technology transfer from those organizations.

On the other hand, R&D organizations can direct their resources to the research activities, without the need to involve themselves in those tasks more close to the market/application. Their involvement in supporting activities like training, consultancy and dissemination can also be considered.

When looking at large companies, several implementation examples of this scenario can be found, since they have to necessary internal resources to cope with the related complexity (people, money, expertise, etc.). But what happens with SME's?

Since they don't have the necessary means to follow the same strategy, SME's have been trying to find alternative ways to address Innovation. Although most of them recognize the need to cooperate with other organizations, the first approach was to simplify to process by "cutting" some of the functions (minimizing the need for other partners and thus the complexity of network management). Three examples are presented to illustrate the consequences of this strategy.

Figure 4 represents the case of projects involving only R&D organizations and end users. This is the typical example of research projects where pilot companies are involved in specification, test and validation tasks. Normally, the end result of these projects are prototypes. The lack of companies capable to transform project results into products and services and to explore them commercially, contributes significantly for the failure of the innovation process.

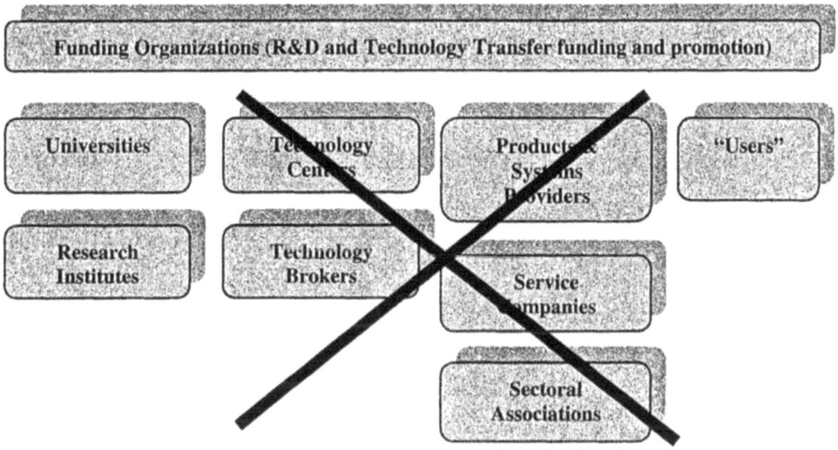

Figure 4 – Example of Innovation process simplification

Another example is the non-participation of end user (projects involve only R&D organizations and product development companies). In many cases, it is very difficult for these SME's to have a real perception of market/customer needs, the consequence being the fact that frequently the result of these projects are products without a real market interest.

Finally, we consider the case of non-participation of research organizations (projects involve technology providers and end-users). The capability of these partners to generate radical innovation and to ensure its continuous update is very reduced, which means that the result is usually incremental and not sustainable Innovation process.

5.2 Strategic Cooperation Networks

The analysis made in the previous sections supports the following statements:

- Most of these functions are crucial for the success of innovation processes.
- Most SME's can't afford to implement and/or support them internally

These facts justify the need to find alternative strategies for the Innovation process implementation and management [2]. Strategic cooperation networks and alliances are emerging as a valuable option for SME's to build Innovation successfully. It is important to mention that also large companies are using this strategy to reduce costs and gain speed and flexibility.

On the other had, the creation and management of these networks and alliances bring new problems and challenges. The following list summarizes our main findings related with this theme.

1. Due to the fact that they are often overlooked, and that leads usually to poor results, the following functions need special attention:

   - Funding: it is necessary to ensure an integrated funding scheme, capable of covering all stages of the innovation process and not only some of them. This usually means the need to integrate different funding programs and funding proposals.
     The industry prototype or market pre-series stages may require venture capital funding, as they are not eligible in pre-competitive research programs. New funding partners/shareholders may enter the consortium at this stage.
   - Global coordination: global coordination and project management is a crucial task to ensure that project runs "smoothly", meaning that each stage performs as planned and mainly that transitions between stages are done in an effective way. This function should be performed preferably by an "independent" organisation (not involved in the "operational" tasks and not committed to any of the partners).

2. Since Innovation implies the commercial exploitation of project results, technology "absorbers/vendors" play a major role in Innovation projects success. Their participation from the beginning or from early stages of the project is crucial. When technology vendors are only involved at the end of the R&D stage, the technology transfer task is more difficult and less efficient.

   Their knowledge of market/end-users requirements is also wasted if they only join the project at later stages.

3. R&D institutions (Universities or Institutes) are usually very important to ensure that project results have a significant degree of Innovation, thus higher potential value and impact, although sometimes at the expense of an higher risk.

4. End users are important to specify and validate project results, although technology vendors can also play this role whenever they have a deep knowledge about the market and customers. The creation of pilot projects/demo sites involving end users can be a major and very effective promotion tool for sectoral dissemination.

6. CONCLUSIONS

The work undertaken so far allows us to derive some conclusions, from which we selected the ones presented below.

The task of managing and promoting innovation projects encompasses a big challenge, and it requires the coordination of several roles/functions and different consortia players, sometimes over long periods of time.

The attempts to simplify the Innovation process by cutting down some of the inherent functions or by assigning some functions to organizations without the necessary skills often led to failure (in terms of Innovation results).

The above proves the need for the development of strategic cooperation networks and alliances, involving different types of companies and institutions, aiming to gather the competences and capacities, which are necessary to create Innovation in a successful and sustainable way. These networks should support long-term relationships, capable of lasting beyond the project execution and of ensuring sustainable Innovation strategies.

In such type of networks, where rather different organisations are involved, each partner has different strengths and assets and it is necessary to deal with the complex task of matching the various, and often conflicting, interests of all parts. This means that the coordination role assumes a critical importance.

7. ACKNOWLEDGEMENTS

The author would like to acknowledge the encouragement and financial support from the Observatório da Ciência e do Ensino Superior (OCES) of the Ministério da Ciência e do Ensino Superior (MCES).

8. REFERENCES

1. Ellis, Caroline – "Making Strategic Alliances Succeed", Harvard Business Review, 1996
2. Caldeira, José Carlos; Chituc, Claudia-Melania; Mendonça, José Manuel - "Strategic Alliances and Innovation Projects Success", The ISPIM 2003 Conference, 2003.

9. BIBLIOGRAPHY AND RELATED INTERNET SITES

1. Ball, Mary Alice; Payne, C. Shirley - "Strategic Alliances: Building Strong Ones and Making Them Last", 1998
2. Ellis, Caroline – "Making Strategic Alliances Succeed", Harvard Business Review, 1996
3. Holbrook, J. A. D.; Padmore, D.; Hughes, L.; Finch, J. - "Innovation in Enterprises in a non-Metropolitan Area: Quantitative and Qualitative Perspectives", 1999
4. OECD, "Oslo Manual", 2nd Edition, DSTI, OECD, Paris 1996
5. OECD "Frascati Manual" or "Proposed Standard Practices for Surveys on Research and Experimental Development", DSTI, OECD, Paris, 2002
6. www.cordis.lu (Community Research & Development Innovation Service)

7. www.fct.mct.pt (Fundação para a Ciência e a Tecnologia, Ministério da Ciência e do Ensino Superior, Portugal)
8. www.adi.pt (Agência de Inovação, Portugal)

PART TWO
ADVANCED LOGISTICS

PLANNING, DESIGN AND MANAGEMENT OF SHARED INFORMATION WITHIN GLOBALLY DISTRIBUTED MANUFACTURING NETWORKS

Tomaso Forzi[1], Peter Laing[2]
Research Institute for Operations Management (FIR)
at Aachen University of Technology (RWTH Aachen)
- E-Business Engineering Department -
e-mail[1]: Tomaso.Forzi@fir.rwth-aachen.de
e-mail[2]: Peter.Laing@fir.rwth-aachen.de

Nowadays, the Internet and web-based E-Business solutions play a crucial enabling role for the design and implementation of new Business Models. This implies high chances, but also remarkable risks for enterprises that choose to pursue a new Business Model or to adapt an existing one. In fact, the implementation of a strategically not appropriate Business Model would critically undermine the long-term success of a company. Hence the clear need for action in the field of methodical Business Modelling. We present a new approach for a customer-oriented (E-) Business Modelling, with a specific attention on inter-organisational co-operative networks and re-intermediation, as well as on information management within distributed manufacturing networks. The approach had been validated for a collaboration network in the German manufacturing industry.

Keywords: E-Business Engineering, Business Modelling, Information Management, Manufacturing Networks, Re-intermediation.

1. NEW ICTs AND THE ERA OF GLOBAL NETWORKS

During the past years, the fast development of new Information and Communication Technologies (ICTs) has been revolutionising the market arena and it extended the horizon of competition (Hagel & Singer, 1999). Such development has been causing radical changes in all business branches (Timmers, 2000). New ICTs and the related new business approaches allow enterprises to design leaner *intra*-organisational processes, with the result of enhancing higher efficiency and productivity (Wirtz, 2000). Besides that, ICTs can strongly influence *inter*-organisational processes, by supporting *co-operation* within *entrepreneurial networks* as well as by enabling their *co-ordination* – by means of Internet-based Business Collaboration Infrastructures (BCIs), such as Electronic Marketplaces, Exchange and Communication Platforms or Supply Chain Integrators (Hoeck & Bleck, 2001). Hence, web-based ICT solutions play a crucial *enabling role* both for new Business

Models and for those models that, up until now, had a higher value on a theoretical than on a practical level (e.g. Virtual Organisations) (Picot et al., 2001). Anyhow, a lot of open issues and unsolved problems in the field of technology fulfilment are still to be faced in order to ease an efficient entrepreneurial collaboration within globally dispersed networks, as e.g. the definition of suitable branch-specific standards for inter-organisational business processes.

As far as inter-organisational processes are concerned, recent trends have shown that companies, in order to face the quick-paced globalisation process, tend to concentrate on the own core competencies, before starting to co-operate within global networks of enterprises (Hagel & Singer, 1999). This transformation process is crucial for the success of the company's business (Picot et al., 2001). Therefore, before making a strategic decision regarding the participation to a co-operation network, the management of an enterprise has to take into consideration a wide set of aspects, such as the entrepreneurial organisation, the own core competencies, the needs of the customers, the readiness of the potential partners to co-operate, and the availability of the needed technologies and tools (Hagel & Singer, 1999) (Porter, 2001) (Timmers, 2000).

In particular, enterprises have to weigh a set of crucial success factors in the fields of standardisation and technology management. For instance, on the one side *communication standards* are essential to achieve an efficient exchange of information and data. On the other side, *branch-specific collaboration standards* are crucial to establish inter-organisational workflows. Furthermore, the *analysis of the potentials of new technologies* can help to select the most suitable technologies and tools for the support of new entrepreneurial businesses.

As a result, in order to deploy innovative *network-oriented co-operation structures*, new (Internet-based) Business Models can be designed and successfully implemented (Afuah & Tucci, 2001) (Rayport, 1999). This includes a process of re-intermediation.

2. ENTREPRENEURIAL STRATEGY AND THE INTERNET: NEW BUSINESS MODELS

In the management literature there are different and discrepant definitions of the term *Business Model* (Afuah & Tucci, 2001) (Timmers, 2000) (Wirtz, 2000). Furthermore, (E)-Business Models can be classified according to a variety of different criteria, such as the degree of innovation and functional integration, the collaboration focus or the involved actors (Rentmeister & Klein, 2001) (Timmers, 2000) (Wirtz, 2000).

The core objective of each company is the long-term creation of *added value* (Porter, 2001). The entrepreneurial *strategy* defines how such a target has to be fulfilled. A successful strategic positioning and a sustained profitability are achieved through an own value proposition, a distinctive value chain, an entrepreneurial fit, and continuity of strategic direction (Porter, 2001). According to our understanding, a *Business Model is an instantiation of an entrepreneurial strategy related to a specific business* and it encompasses: market (customers and competitors), outputs, revenues, production design, partner network and financing (see Figure 1).

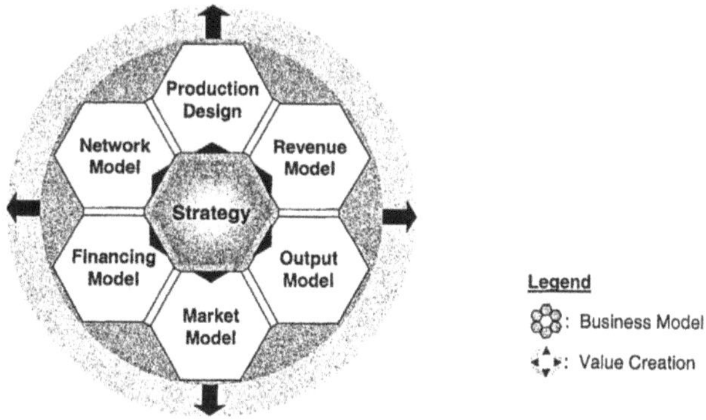

Figure 1 – Strategy, Business Model and creation of Added Value

In the new ICT era, companies strive to exploit the advantages of the networked economy and to dodge the related risks. Lately, Business Models tend to involve different enterprises with the goal of bringing higher profits to each of the participants (Afuah & Tucci, 2001) (Picot et al., 2001). The modelling of such co-operative E-Business Models represents a significant challenge, since there are no appropriate methods to tackle systematically such modelling issue. *We developed a customer-oriented business modelling approach that includes the design of information management within dispersed manufacturing networks.*

3. CUSTOMER-ORIENTED BUSINESS MODELLING

Due to the globalisation, the growing transparency of the markets and the resulting increased competition, enterprises have to focus more and more on their customers. As a matter of fact, the fulfilment of the customers' needs is an essential precondition to generate a sustainable turnover. This means that enterprises have to take this aspect into deep consideration and must consequently shape Business Models that reflect the customers' needs.

Up until now the development and adjustment of Business Models has been performed by companies mostly in a creative way (see Chapter 2). As a matter of fact, in the state-of-the-art there is hardly any available methodical support (Afuah & Tucci, 2001) (Picot et al., 2001) (Rayport, 1999). We are therefore convinced that a successful approach to tackle this methodical lack must be based on a strategic focus on the *customers' needs*. Within a running research project, we have been developing the *House of Value Creation (HVC),* a method to design customer-oriented and sustainable Business Models (Laing & Forzi, 2001) (see Figure 2). The HVC is a meta-method, since it consists of *three logical pillars* (input, method, and output) and of *six process layers* (each of the process steps requires a suitable method). The method suits explicitly the design of Internet-based and network-oriented Business Models.

Because of our understanding of the term Business Model and because of the fact that the ultimate goal of a company is long-term value creation, the design of a

Business Model has to be definitely based upon the entrepreneurial strategy. Therefore, if not already done, the first step within a business modelling process is to define the rough entrepreneurial strategy, on which the Business Model will be based. The meta-method of the House of Value Creation illustrates the correlation between a set of significant leverages (first pillar of the HVC), the Customer-oriented Business Modelling process (second pillar), and the resulting Business Model (third pillar). As previously hinted, our Business Modelling approach encompasses six layers that correspond to six following steps of the method. The first HVC phase is triggered either by the inside or by the outside of the company – through a new (business) idea, invention, innovation or modifications of the economical environment. The six steps of the method are presented in the next sections.

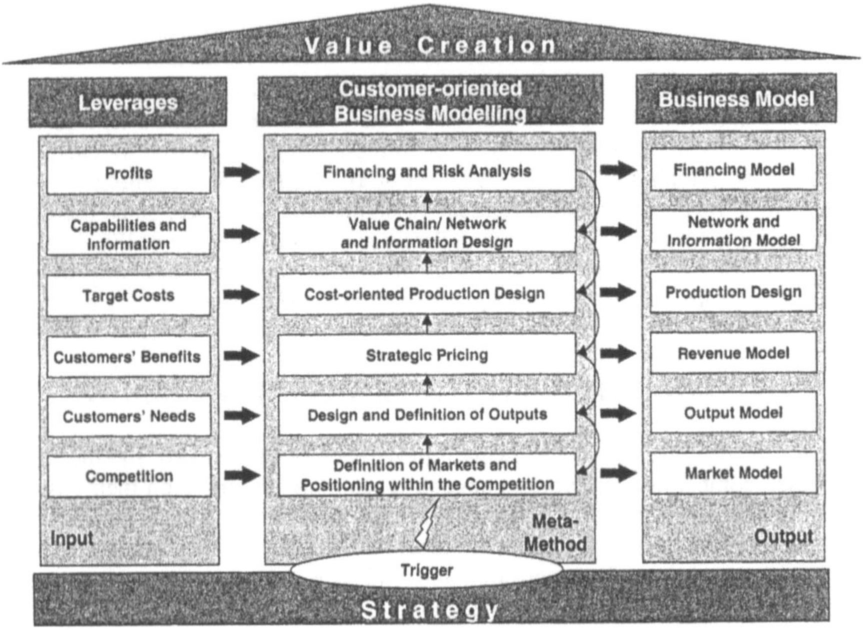

Figure 2 – The House of Value Creation

3.1 Definition of the markets and positioning within the competition

The initial decision regards the category of products or services to deal with (as well as the substitutes). Hence, a consequent monitoring of all strong players (suppliers, customers, and possible competitors) has to be done. This phase deals with the branch profitability and with the rivalry among existing and potential competitors. The current as well as the expected branch profitability must be deeply taken into consideration. Phase output: *market model*, with a clear identification of the key players, customers and competitors.

3.2 Definition and design of the outputs

In the product design, a well-proven method is the Quality Function Deployment (QFD) (Akao, 1990). With this approach, a customer-oriented product development can be successfully realised. Therefore, after defining markets and identifying core customers and competitors, the outputs (products or services) have to be shaped in order to maximise the customers' benefit according to QFD-like method. Phase output: *output model*, with a detailed customer-oriented design of the striven products and/ or services.

3.3 Strategic pricing

The identification of prices for the planned outputs should be more the result of a strategic positioning than of a cost-oriented approach (Hagel & Singer, 1999). The price calculation should take into consideration the customers' surplus constraint as well as the strength of the competition (existing barriers of entry, such as patents, industry property rights, etc.) (Kim & Mauborgne, 2000). Phase output: *revenue model*, with a detailed description of how earnings can be achieved.

3.4 Cost-oriented production design

According to the guidelines of the revenue model, the target costs for the output model will be calculated (as the upper bound for direct costs). Hence, the requirements to the value chain will be detailed. Phase output: *production design*, with a detailed description of how the performances have to be achieved.

3.5 Partners, network and information

In this phase, starting from the requirements on performances of the value chain, the capabilities (core competencies, capacities, available modules and components, ICT infrastructure) of the own performance structure and of the potential partners will be thoroughly scanned. Phase output: *network model*, selection of the partners within a specific instantiation of the value chain and network configuration, as well as *information model*, i.e. the approach according to which the information management issue has to be tackled.

3.6 Financing and risk analysis

Eventually, based upon the expected profits and a suitable scenario analysis, the risk-level as well as the need for working capital must be calculated to start the search for investors (Fink et al., 2000). Phase output: *financing model*.

At each step of the HVC the corresponding targets must be fulfilled. If one step is not fulfilled, then the process should go back to a prior phase as long as the issue is tackled – with an iterative approach. In the following chapter, our contribution will focus on a specific level of our HVC, namely the one regarding *network* and *information model*. A specific attention will be paid to the necessity to manage information and the related information flows within manufacturing networks, as well as to the approach to tackle the issue.

4. THE DESIGN OF INFORMATION MODELS IN A NETWORKED ECONOMY

In the fifth phase of the HVC customer-oriented Business Modelling, starting from the requirements on performances of the value chain, the state-of-the-art and the capabilities of the own performance and information structures, as well as of the ones of the potential partners, will be thoroughly scanned. The striven value chain will be designed and a specific network will be instantiated. (see Figure 3).

Figure 3 – The fifth level of the House of Value Creation

As stated in the initial section, before making a strategic decision regarding the creation of a co-operation network, the management of an enterprise has to take into consideration a wide set of aspects, such as the entrepreneurial organisation, the own core competencies and the ones of the partners, and the available technology. The gathering of all information regarding the potential network participants represents what we define as *capabilities and information*. This is one of the tasks of the network broker. According to the requirements previously identified within the *production design* (output of phase 4 of the HVC), a crosscheck with the capabilities of all potential partners helps to define a set of suitable partners. This process can be supported e.g. by the use of specific broadcasting platforms such as E-Marketplaces. The data about the entrepreneurial capabilities must regard both the own core competencies and the ICT capabilities (Krcmar, 2000) (Picot et al., 2001). As a result, the instantiation of a *specific network configuration* can be identified. Furthermore, the definition of inter-organisational blue prints and branch-specific process standards is necessary to enable lean and efficient *inter-organisational processes and workflows*. The result is the *network model* (see Figure 4).

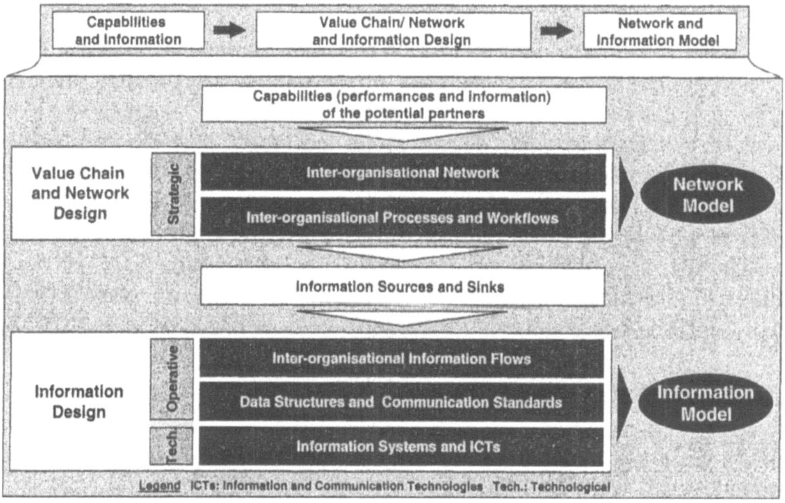

Figure 4 – Inter-organisational Network and Information Models

Within the present contribution, we consider the network model as given: we assume that the value chain had been designed and that a specific network configuration had been identified. The most relevant aspect for the following phase is that the inter-organisational processes determine the *sources* and the *sinks* of distributed information.

The *information design* focuses on the management of the information within such entrepreneurial networks and inter-organisational value chains (see Figure 4). Within the modelling of an inter-organisational information management model, we distinguish three different layers: information flows, data structures and communication standards, information systems and ICTs.

4.1 Inter-organisational Information Flows

Starting from the sources and the sinks of distributed information, the inter-organisational information flows can be derived and consequently defined. Within this process, several different aspects have to be taken into consideration: *direction* (e.g. one way or bi-directional information flow), involved *actors* (broker, network members, and related intelligent agents), involved *systems* (e.g. knowledge management, distributed Data Base information sources), and *relationship* (e.g. 1:1, 1:n, m:n). Furthermore, information flows encompass two other major dimensions: *information width*, which sets the different degrees of information broadcasting and transparency within the network, and *information depth*, which defines the scope, aggregation level and content of the information flow. Eventually, the modelling of information flows in relation of complexity the considered job and of the media richness (regarding e.g. real-time or asynchronous response) can be done according to the *Media Richness Theory* (Daft & Lengel, 1984).

4.2 Inter-organisational Data Structures, Communication Standards

A crucial issue is the development of an inter-organizational *data model* to ensure the consistency, quality and transparency of the data. A data model encompasses also the aggregation levels for Management Support Systems (e.g. Data Warehousing and Data Mining). Furthermore, the branch-specific development of *shared communication standards* (e.g. based on XML) is essential to achieve and ensure a fast exchange of information and data through inter-organisational networks, as well as the integration of different ERP systems within an Internet-based BCI. The identification of the most appropriate technology standards is important also to deploy a back-end integration in each of the involved enterprises. Eventually, an appropriate solution for the *security issue* has to be identified.

4.3 Information Systems and ICTs.

Enterprises have to weigh a set of crucial success factors in the field of the technology management, since the analysis of the potentials of new technologies can help them to select the most suitable technologies and tools to face challenges in the context of new collaborative businesses. As a matter of fact, a proper *technology and innovation management* is important to plan the development of the entrepreneurial core businesses in co-ordination with a strategic technology planning (Krcmar, 2000). Last but not least, a *make or buy* decision has to be made, e.g. by exploiting the services of an Application Service Provider.

5. CASE STUDY: INFORMATION MANAGEMENT WITHIN MANUFACTURING NETWORKS

Each industrial sector has its own peculiarities; therefore, it is necessary to develop approaches and solutions that take into account such specific requirements. Our institution has a proved experience with German SMEs of the manufacturing and machinery industry. Since we identified a remarkable *need for "e-action"* in this traditional and static industrial branch, we therefore decided to develop *customized Business Models and E-Business solutions* with a high impact onto this industrial sector. In this context we developed the case study for the present contribution.

The presented approach has been validated in the case of the development of a Business Model for an intermediary for the metal processing and manufacturing industries. The two year project, founded by the German Federal Ministry of Economy and Technology (grant number: VI B 4 – 00 30 60/35), started in 2001 with a consortium of 10 SMEs of the above-mentioned industrial sectors. The project is named "Z-Online", where "Z" stands for the German word *"Zeugnis"* which means "certificate". Hence, the project name means "Certificate-Online".

Z-Online focuses on the field of *manufacturing networks for metallic material.* As a matter of fact, the generation and mailing of so-called *paper-based test reports*, which document and guarantee to the buyer specific properties of the metallic material, is nowadays accompanied by several serious problems. For instance, the open issues concern the archiving of reports and the specification check of corresponding material standards referenced in a test report, as specified by DIN, the German Institute for Standardization (DIN 10 204, 1990).

The innovative *trigger* for Z-Online is the worldwide dissemination and acceptance of the Internet as communication and information exchange channel as well as the idea to exchange material test reports electronically. This makes the development of new intermediary services for the efficient exchange and storage of material test reports based on an electronic Business Collaboration Infrastructure feasible. Based upon the innovative idea of exchanging electronic material reports on a collaborative Internet-platform, we proposed a Business Model – the result of the use of the House of Value Creation. We are now going to present the results obtained within the business modelling process.

5.1 Market model

After a thorough market analysis we identified similar solutions to manage and exchange material reports (e.g. Document Management Systems, DMS), but we realized that none of them fulfils all relevant requirements. Hence, in the resulting market model there are *no direct competitors* because of the fact that the planned service is innovative and therefore it is not offered in the market yet. Furthermore, the *potential customers* are all the manufacturing enterprises that exchange metallic products – i.e. both ferrous metals (all sorts of steel) and non-ferrous metals (such as tin, copper, and brass) - with specified and guaranteed properties, as well the different testing and inspection organizations (such as TÜV or Dekra in Germany, Det Norske Veritas in Norway, BSI or ITS in the United Kingdom, SGS in Switzlerland, Bureau Veritas in France).

5.2 Output model

We conducted several workshops with the involved SMEs and thus we gathered all the requirements related to the exchange of electronic material test reports in the manufacturing field. We distinguish two kinds of groups of services required by the potential customers of the platform:

- *Basic service,* e.g. storage, access, and remote archiving of electronic material test reports over the Internet.
- *Value-added services,* e.g. assessment and evaluation of the suppliers, nominal/actual value comparison of measures certified in a test report, batch management, offering of detailed information about material-related quality.

Other general requirements on the overall performances of the platform are: a high operating efficiency and flexibility, specific security requirements (e.g. transmission and privacy), a 24x7 hours system availability, a suitable multi user concept with different and adjustable levels of authorizations, support of surveyors of independent testing and inspection organizations.

5.3 Revenue model

After an analysis of the customers' benefit, we used a two-step *price corridor method* for the strategic pricing for shaping a revenue model (Forzi & Laing, 2002):

a) Identification of the price corridor of the mass, i.e. search for the price corridor that the majority of the customers is willing to bear. According to the market model, there are no direct competitors, but only some possible substitutes, i.e. providers of DMS, whose products, though, do not fulfil all relevant customers' requirements. We observed that the innovative services of the planned intermediary service infrastructure might therefore crucially change the power balance in the market of tools for the management and exchange of material test reports. The current cost to process a single material report amounts up to about 50 €, which clearly represents an upper bound for the price model. Since the process cost for an electronic test report drops drastically by the use of the intermediary service infrastructure, the decision was to pursue a *low* price corridor strategy to target a high number of customers.

b) Specification of a level within the price corridor, i.e. identification of an appropriate price level within the chosen low price corridor. A detailed analysis of the customers' benefit of the DMS underlined that none of them can fulfil all industrial requirements. Therefore, the intermediary service infrastructure with its innovative customer-oriented services has realistic chances to be widely accepted by the target group and thus to penetrate successfully the market. In order to conquer the market and achieve the striven critical mass in terms of traffic (reports/period of time), it was furthermore decided to choose a lower pricing-level within the chosen low price corridor. A high traffic, though, does not yet guarantee a long-term success, because second-movers might come up with similar solutions and gain quickly market share. Hence, the price should be maintained very low until the critical mass in terms of branch members is also reached. With the achievement of this goal, the developed *format for electronic test reports* will be widely disseminated and it has therefore good chances to be accepted and adopted as a branch-specific standard. At this stage, barriers for market entry for possible competitors will be significant. The deployment of a mid or upper-level pricing within the selected price corridor will be then possible without risking to loose market shares. A potential cash cow for the business is represented by the portfolio of attractive value-added services.

5.4 Production design

According to the guidelines of the revenue model, the target costs for the output model will be calculated within the *cost-oriented production design*, i.e. design of the electronic transmission, management and storage of material reports. In the case of the "production" design for the transmission of material reports, the attention was paid to the fixed costs (i.e. target costs for the infrastructure) since direct costs (i.e. cost for the report transmission) tend to zero. Hence, the result is a platform with a targeted low fixed cost (e.g. hardware, software and mainly personnel costs). The lower the fixed costs are, the sooner the critical mass in terms of participants and transactions will be reached.

As far as concerns the design of the platform, we identified the need for the following capabilities: database based material test reports management and archiving, material science know how, trust management, and information content for value-added services.

5.5 Network and information model

In this phase, starting from the requirements on performances and from the capabilities, core competencies and IT infrastructure of the involved potential partners, a specific performance network as well as an information and technology framework were defined.

The targeted market consists of several SMEs, of which none of them is dominating the market. It is important that all enterprises that take part to the platform must trust and be able to rely on the carrier. Hence, the managing institution of the transaction platform for electronic test reports must be an independent company. It was thus decided that the collaboration platform should *not* be managed by one of the manufacturers of metallic material, but by a neutral intermediary with material science know-how as well as ICT competence (e.g. data base management and archiving). As a matter of fact, the archive management and value-added services that require particular material science know-how will be performed by the intermediary, who will outsource the other competencies to two different partners; one partner was identified to deal with trust management and another to provide information content for the other value-added services of the platform.

Within the analysis of the information infrastructure, the crucial issue was the modelling of processes and the management of shared information to enable the *inter*-organisational and *intra*-organisational workflow capabilities of the planned ICT-System. Furthermore, some of the other most interesting aspects that were dealt with are the definition of *process standards* for inter-organisational processes and workflows, the development of an appropriate and flexible interface to deploy a *back-end integration* in each of the involved enterprises, and the branch-specific development of a *shared standard* for electronic test reports. Such standards ensure a fast exchange of the required information through inter-organisational networks as well as the integration of different ERP systems.

5.6 Financing model

Because of clear privacy reasons, we are at the moment not allowed to distribute information about the financing model.

6. CONCLUSIVE CONSIDERATIONS

The trend towards a tightly inter-connected economy seems to be nowadays unquestionable. On the other side, the development process towards a dynamically networked economy has just recently started. We strongly believe that, in order to be successful within a networked economy, enterprises will thus have to undergo a deep transformation process in their organisational philosophy, in their structure, in the used methods as well as in their approach to interact with external organizations. We are convinced that, in order to face the challenges of such a dynamic and insecure business context, an appropriate business modelling approach and in particular targeted solutions to manage networked information are key success factors to plan and hence deploy sustainable and dynamic businesses.

7. REFERENCES

1. Afuah A, Tucci CL. Internet Business Models and Strategies: Text and Cases. New York: McGraw-Hill/Irwin, 2001.
2. Akao Y. Quality Function Deployment: Integrating Customer Requirements into Product Design. Portland: Productivity Press, 1990.
3. Daft R, Lengel R. Information Richness: A New Approach to Managerial Behaviour and Organization Design. Research in Organizational Behaviour 1984 (6): 191-233.
4. DIN (publisher) 10 204. Metallic products – Types of inspection documents (German version). Beuth Verlag, 1990.
5. Fink A, Schlake O, Siebe A. „Wie Sie mit Szenarien die Zukunft vorausdenken". Harvard Business Manager 2000 (2): 34-47.
6. Forzi T, Laing P. Business Modeling for E-Collaboration Networks. In: Proceedings of the 2002 Information Resources Management Association International Conference (IRMA 2002), Seattle, May 19th- 22nd, 2002: 961-963.
7. Hagel III J, Singer M. „Unbundling the Corporation". Harvard Business Review 1999, 77(2): 133-141.
8. Hoeck H, Bleck S. „Electronic Markets for Services". In: E-Work and E-Commerce, volume 1, Stanford-Smith B & Chiozza E editors, Amsterdam Berlin Oxford: IOS Press, 2001, 458-464.
9. Kim C, Mauborgne R. "Knowing a Winning Business Idea When You See One" Harvard Business Review 2000, 78(5): 129-138.
10. Krcmar H. Informationsmanagemnt. 2nd ed. Berlin, Heidelberg, New York: Springer, 2000.
11. Laing P, Forzi T. "E-Business and Entrepreneurial Cooperation". In Proceedings of the 1st Intern. Conference on E-Business (ICEB 2001), Hong Kong, December 19th–21st, 2001: 7-9.
12. Picot A, Reichwald R, Wigand RT. Die grenzenlose Unternehmung – Information, Organisation und Management. 4th ed. Wiesbaden: Gabler Verlag, 2001.
13. Porter ME. "Strategy and the Internet". Harvard Business Review 2001, 79(3): 63-68.
14. Rayport JF. The Truth about Internet Business Models, 1999. Retrieved on September 27th, 2001, from the Web-site: www.strategy-Business.com/briefs/99301/page1.html.
15. Rentmeister J, Klein St. „Geschäftsmodelle in der New Economy". Das Wirtschaftsstudium 2001, 30(3): 354-361.
16. Timmers P. Electronic Commerce. John Wiley & Sons, Ltd, 2000.
17. Wirtz BW. „Electronic Business". Dr. Th. Gabler Verlag, 2000.

CUSTOMER RELATIONSHIP MANAGEMENT AND SMART ORGANIZATION

Peter Weiß
FZI Forschungszentrum Informatik, Germany, pweiss@fzi.de

Bernhard Kölmel
CAS Software AG, Germany, bernhard.koelmel@cas.de

The increasing concentration of an astonishing high number of enterprises regarding their core competencies causes a constant growth of virtualization and distribution of value chains of companies. This happens either by a spin-off in subsidiary companies or by a cross-company co-operation. This phenomenon induces mainly higher requirements concerning the inter company co-operation and the customer relationship management. The concrete implementation and the possibilities of a groupware-based customer relationship management for virtual organizations can be shown by using the Infranet Partners (dynamic network of European SMEs) as an example.

Smart organization, Customer relationship management, Cross-company co-operation, Dynamic networked organization, Best practice pilot

1. INTRODUCTION

The Information Society (IS) offers great opportunities and challenges for European small and medium-sized enterprises (SMEs). SMEs are expected to uptake innovative ICT solutions and e-commerce and innovative new methods of work to take advantage of the IS. *Customer Relationship Management (CRM)* is seen by experts of vitally importance for business networking as it can be evaluated as one of the major drivers to build up smart organizational structures in small businesses. SMEs face a no doubt enormous pressure in the inter-networked economy with increasing internationalization and globalization of markets with more and more difficulty to build up stable and long-term relationships to their customers. Smart organizations are one example for nowadays business networking (Fleisch, Österle 2000) which ranks amongst the most important capabilities businesses will need in the information age. Especially for SMEs this kind of co-operation is seen of high importance and relevance especially because SMEs are the cornerstone of Europe's competitive position and job creation. But without doubt as already mentioned there is a lack of information and best practice how to build up smart organizations. As well as only little knowledge of best practice with regard to success factors and barriers. This situation is the starting point for Infranet. Our best-practice pilot shows us typical problems for SMEs to adopt and apply ICT enabled business networking.

2. SMART ORGANIZATION

Over the past years the steady rise of market globalisation, the increased stress of competition and the shortened product life cycles caused conditions, which need new organizational forms (Pribilla, Reichwald, Goecke, 1994). New organizational forms differ from the traditionally and hierarchically structured ones. The new forms are characterised by„ [...] open and flexible, fluid and easy systems with only little hierarchy"(Picot, Reichwald 1996, pg.10). In order to stay competitive, the companies need to have the necessary flexibility and a fast reaction. For this they must reduce their complexity in order to concentrate on their strengths, which are the core competencies. This can be realised by sourcing out business activities with the aid of information technology. In respect of these developments one can say, that the so far existing classical types of enterprises are not as well-defined as they were. They blur more and more, change steadily and yet vanish partially. Organization visionaries like Malone prognosticate for the organization of the 21st century a scenario, in which small enterprises will act in big networks and will join together for projects in „*smart organizations*".

2.1 Structure of Smart Organization

Organizations structures are called *virtual*, if legally independent enterprise, institutions and/or individuals join together and appear to a third party as one enterprise, which co-operates in order to carry out common business interests. This co-operation is based as well on a steady (*static*) as on a *dynamical network*. Characteristic features are a massive use of information- and communication-technology (ICT) in order to support the internal and inter-company co-ordination and co-operation to compensate the non existing centrally located management functions of the company. The aim of this co-operation is the value chain optimization by bringing out core competencies, as well as the splitting of risks, costs and knowledge of each partner. Another aim is the appearance as one competent and trustworthy unity (Längsfeld, Weiß; 2000). Based on extensive literature research three basic types of virtual organizational structures were defined by Längsfeld/ Weiß. Within an empirical study the existence and relevance in industry of these structures was tested. In the following table/figure 1 and figure 2, information on the structure and goal of the different types is given. Definitions and characteristics used by a representative selection of contributions of twenty (20) authors were condensed to a working definition of smart organizations (see paragraph below). Furthermore, a measurement based on seven dimensions was deduced from literature.

In order to have a common starting point and understanding of all participating respondents the following working definition of Virtual Organizations (VO) was as follows:

Working Definition of Virtual Organization:

Organizational structures are called virtual, if legally independent enterprises institutions and/or individuals, which appear and co-operate with the other(s) as an enterprise, in order to pursue common business interests. Co-operation can be based on either a stable (static) or a dynamic network. A characterising feature is a substantial application of information and communication technology (ICT) for the support of intra- or inter-company co-ordination and co-operation for the compensation of central management functions of the enterprise. Virtual Organizations aim at the optimization of the value creation chain by bringing in core competencies, as well as the spread of risk, costs etc. of the individual partners.

Table 1 – Types of virtual organizations (Längsfeld/ Weiß, 2000)

Type	Description
type I	▪ long term oriented partnership ▪ focal partners dominate ▪ optimisation of the value chain
type II	▪ dynamic project oriented partnership ▪ heterarchical structure ▪ established partnerships
type III	▪ temporary partnership ▪ short term market chances

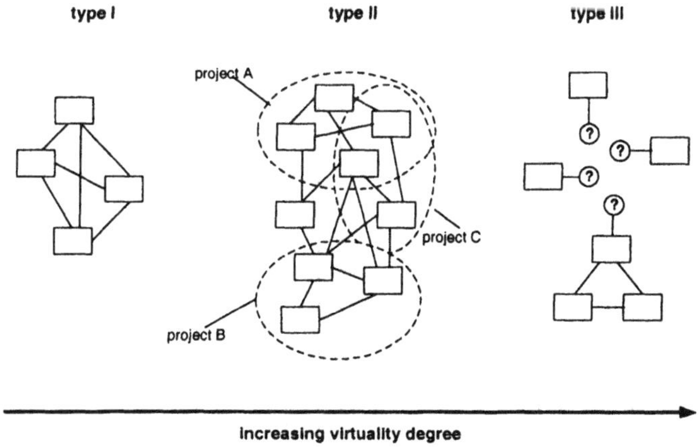

Figure 1 – Degree of virtuality

2.2 Motivation and Goal of the Empirical Study

Since the early nineties, Virtual Organizations (VO) as one of these new organizational models has been in the focus of scientific research. Among the most frequently researched areas is the assistance of VO with adequate tools (IT-tools and modelling tools) and business models. However, there is very little evidence on the degree of take up and state of the art of VO in industry. Therefore, the goal of this study in the year 1999/2000 was to analyse the basic structures of VO and to investigate how these structures are supported by innovative ICT.

Our assumption is that relationships between companies are rather complex and companies in real life business may take different roles simultaneously in different business relationships. This approach is in line with ongoing investigations in business networking assuming that companies at the same time act as suppliers of larger companies, customers/ suppliers of other SMEs, technology providers and solution partners. Consequently, business reality is too complex to be described by means of a single theoretical model. However, a theoretical model enables by simplifying a deeper understanding of the key drivers and criteria to investigate business ecosystems. A methodology often applied in socio-economic research approaches. The applied methodology aims to compare identified networks and co-operations against a continuum framework build on simplified, idealised network types. These types form a continuum of virtuality. The degree of virtuality is measured as already mentioned in seven dimensions derived from a profound assessment of scientific literature (status 1999).

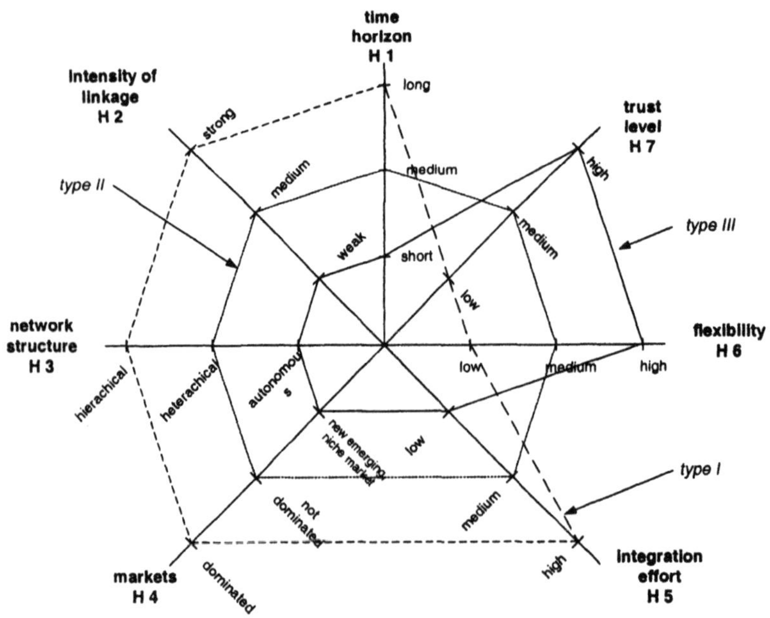

Figure 2 – Seven dimensions to measure virtuality degree

Seven dimensions as shown in figure 2 (flexibility, time-bonding, trust, market, need of integration, intensity of linkage, network structure) were condensed into four coefficients: (1) degree of centralisation, (2) intensity of linkage, (3) flexibility and (4) confidence. Our investigations have been carried out, trying to find structures and similarities that could help to classify virtual companies.

2.3 Results of Cluster Analysis

The analysis of our empirical field study showed that our elaborated virtual organization continuum does adequately and significantly measure virtuality degree. Consequently, our empirical model allows classifying identified real life use cases and distinguishing between the different virtual organizational structures. Our analysis identified four different clusters along the continuum. The resulting classification of use cases and the according distribution is presented in the following figure 3 and in table 2.

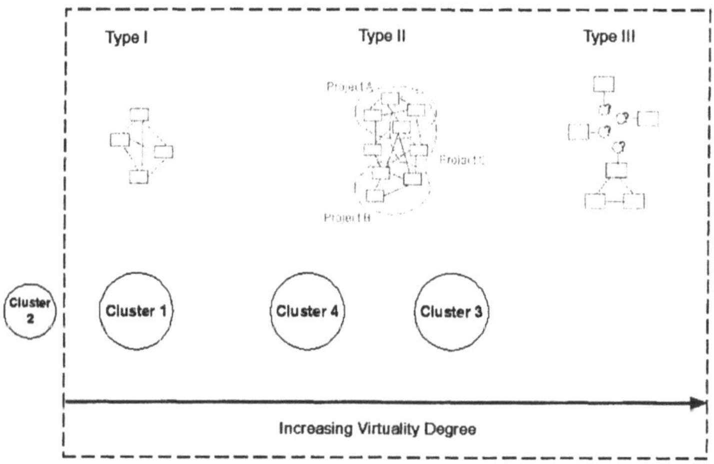

Figure 3 – Clusters of virtual organization along the continuum

Table 2 – Clusters (Längsfeld/ Weiß, 2000)

	Cluster			
	1	2	3	4
Degree of Centralisation	5,2	1,4	2,5	2,7
Intensity of Linkage	5,0	3,0	2,2	4,8
Flexibility	4,8	3,6	4,4	4,9
Confidence/ trust	5,6	2,3	5,6	4,4
Number of cases	***31***	***5***	***27***	***31***

The achieved results shown in above table two (2) are to be interpreted as follows:

Cluster 1

Networks formed by enterprises of cluster one (1) are structured strongly hierarchically. A strong focus on one or several focal business partners dominating the network can be observed. This is a typical constellation that can be found between actors in vertical value creation chains (i.e. supply chains). The intensity of linkage (integration) between these enterprises is of highest degree in comparison to all observed clusters, i.e. the business processes of the network partners are structured across company borders. An interesting observation is that networks represented in this cluster still feel themselves relatively flexible, despite of their hierarchical network structure and a high intensity of linkage. Consequently, we conclude that a high level of integration between the partners obviously does not imply automatically that the network feels inflexible regarding the integration of new partners.

Cluster 2

Enterprises belonging to cluster two (2), indicate the lowest values in almost all applied factors (e.g. degree of centralisation (1,4), flexibility (3,6) and the factor confidence is particularly low in value compared with the other clusters shown in figure 3 and table 2 (2.3 compared with 5.6, 5.6 and 4.4)). Consequently, the entities represented in this cluster are characterised by a strong hierarchy in network structure, large inflexibility while cooperating with business partners and furthermore a extremely low confidence/ trust. Obviously, these companies do not trust their business partners. They want to keep control themselves regarding all business processes and transactions.

Cluster 3

The companies represented by cluster three (3) are controlled in a relatively decentralised way. Nevertheless, the network structure can be significantly characterised as hierarchical. The business processes are weakly integrated between the network entities (business partners) (2,2). The network is characterised by a high level of flexibility and a very high level of confidence/ trust. Conclusively, cluster three can be summarised as a highly flexible, project-centred, dynamic networked, not centralised business network.

Cluster 4

The networks represented by cluster four (4) are matching with those in the cluster before. The companies show significantly the same characteristics. We observe as only significant difference a relatively high value for intensity of linkage (4,8) compared to the cluster before. The level of confidence/ trust in the network is slightly lower (4,4) than the values in clusters one and three. Consequently, networks represented in this cluster are described as flexible and hierarchically structured. The business processes and tasks within the business network are significantly interdependent and strongly interlaced. Therefore, this cluster is characterised by a lower virtuality degree in comparison to cluster 3 and is positioned in the introduced continuum between type I and II.

2.4 Cross-business Process View

A cross-business process view is necessary in order to achieve more flexibility against the background of the increasing complexity of inter-company interfaces. The designing of these processes requires on the one hand concepts based on business economics. These concepts stay abreast of cross-company approaches. On the other hand user-friendly information- and customer-relationship-management is essential for the fulfilment of the design and for the time flow control. A multilevel based empirical research of the Swiss economy on the co-operation topic in the 80ies showed, that co-operation takes place especially on the distribution level and on that concerning the product programme. Similar to this is the division of work with regard to production. According to Endres the furthest development of co-operation is carried out in the field of purchasing, distribution, marketing, market research and information technology. "Here we see a quick success in contrast to the co-operation e.g. on the field of product development" (Endres, 1991, p.25). The results of the studies show, that structures of virtual organizations are set up in order to assure or to meliorate the own competitive position. The communication and relation to the customer is vitally important for the success regarding a co-operation in the field of distribution and marketing. Consequently a successful Customer Relationship Management (CRM) is absolutely necessary for reaching the wanted co-operation and therefore for realizing a co-operation on the basis of structures of VO.

3. IMPORTANCE OF CUSTOMER RELATIONSHIP MANAGEMENT IN VIRTUAL ORGANIZATIONS

Concerning smart organizations[i] *Customer Relationship Management (CRM)* is understood as a concept for designing relationships among an enterprise network and its customers. CRM covers a systematic acquisition, maintenance and intensification of long-term relationships. It is the aim of CRM on the one hand to maximize *Customer Lifetime Value* for enterprise purposes and on the other hand to achieve a high level of customer satisfaction. According to Henry Ford "selling a car [...] is not a business deal, but the beginning of a relationship". For smart organizations CRM is a holistic approach for specific business management. CRM does integrate and optimise cross-company processes with respect to marketing, distribution, customer service, invoice issuing and development. In general this happens on the basis of a CRM software and on a CRM process defined before. According to Lindner (Lindner, 2001) the CRM system is a crucial mean for putting cross-company concepts into practice and he also says that smart organizations will face further problems. Regarding the introduction of CRM a specific strategy is necessary, which must take the following preconditions into account:

- Small division of work in virtual enterprises. In SMEs often telephone switchboards are met instead of call centres. In the IT-industry technical requests are often answered by the product development, which frequently is also responsible for the helpdesk and the presales. In many cases there is only one contact person per customer.

- Small standardization. The small division of work in small enterprises reduces the pressure on the companies themselves and therefore also on virtual organizations to standardize the course of business and to adhere strictly to it. From one case to the other the employees perform a lot of tasks with common sense. The prioritisation depends also on the excitement of a customer. The level of system and professionalism depends on the view of work of the responsible employee.
- Because of dynamic of a virtual organization is given, the new network is affiliated and integrated quickly. The fast activation and set up as well as the easy disconnection of communication relationships must be supported efficiently (e.g. particularly in form of interoperability among heterogeneous IT-architectures) (Mertens, 1998, pp.68).
- The integration of mostly heterogeneous information systems is a big problem in co-operation projects. In this connection costly co-ordination processes and time-consuming interfaces definition are a hindrance factor. This circumstance becomes visible in a historically grown enterprise, where several isolated applications are linked with isolated data. The arising problems appear accordingly in a cross-company integration (Faisst, 1997, p.4).

4. INFRANET PARTNERS – A PRACTICAL EXAMPLE

The Infranet Partners are an example for the international Virtual Organization (see also: http://infranet-partners.com/). The Infranet Partners form a competency co-operation of SMEs for "[...] providing innovative, interoperable end-to-end solutions, adding value to our customers by pooling our resources and expertise [...]". The focal point of this co-operation is an increased flexibility by bringing together resources and experiences, as well as the sort and quality of products and services towards the customer. The Infranet Partners have a specific mission in their co-operation. They represent a steady and *stable relation business network*. With the aid of their business network common projects are realized. The co-operation among the partners is intended to be for a long-term-time and the network partners strike for a common goal: serving common clients Europe-wide with customized products and services. The business network does match in structure with type II in the continuum of virtuality (Längsfeld/ Weiß 2000) already described in section 2.1.

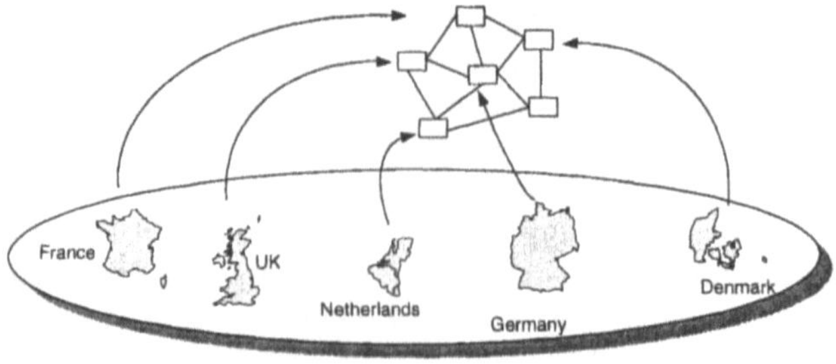

Figure 4 – Infranet partner network

4.1 Type II of Virtual Organization

A network of this character can be described as flexible and non-hierarchically structured enterprise network, in which *business processes* are *seldom* interlaced with the process of the network partners. The enterprises are controlled in a relatively decentralised way. There is a high flexibility and a very high confidence between the partners. This results in a very flexible, not centralised partner network.

4.2 Services and Products

Thus the Infranet Partners have the possibility to offer their services and products even if there have to compete with big multinational companies. The business understanding among the network partners is based on confidence without setting out arrangements in writing. The companies meet monthly in order to discuss current problems and further strategies as well as to strengthen personal relationships. According to Infranet Partners it would be the end of the co-operation, if one would seek legal advice for arising problems. The companies quest for an optimal *value chain* with the aid of their core competencies. If necessary these core competencies are combined in order to carry out common projects. The Infranet Partners appear towards a third-party as one corporate identity. This is a very important characteristic for the partners. A common internet page presents all partner companies as a common brand (http://www.infranet-partners.com). The network slogan is: "Each country, one partner" and it emphasizes the already existing "corporate identity". Up to now there a no central management functions in the network of the Infranet Partners. All important decisions are taken on the common partner meetings. At the moment a kind of "central office" is responsible for the organization of common meetings in turn with different partners. This office also writes and hands out minutes, documents and information. In so doing the "central office" co-ordinates the network. This kind of network can succeed only with the aid of a massive and efficient use of information- and communication-technology in support of internal and inter company co-operation and co-ordination towards the customer. The project called Infranet (see also: http://www.infranet-web.org) under the direction of the FZI adapts a suitable ICT solution to the needs of the smart organization. The identified weak points are supported by qualified ICT-solutions. Because of the requirements and the mission of the network the CRM-groupware genesisWorld of the CAS Software AG in Karlsruhe, Germany was chosen and implemented. The software-solution supports particularly cross-company business processes in the network and boost their efficiency. In many cases the central aim of nowadays co-operation namely the improvement of distribution and marketing as well as the relation to customers is neglected. In this lies the great challenge for the success of virtual organization structures. The ability, to service customers quickly and extensively – in other words a successful Customer Relationship Management – is for this kind of co-operation the road to success.

5. REALISATION OF CUSTOMER RELATIONSHIP MANAGEMENT IN A VIRTUAL ORGANIZATION

Because of the interlocking of each network members of SMEs the used CRM-solution must show the most important capability characteristics of a groupware-system (e. g. address-, date- and document-management). The CRM-solution must also co-operate with merchandise information systems of medium-sized companies. An isolated CRM makes no sense:

- An appointment with a customer concerns a CRM- and groupware-system.
- The customer file comprises offers from the enterprise resource planning system as well as letters of complaint from the groupware document-management.
- The customer advisor must access all data from CRM-, merchandise- and groupware-systems without application change. Interfaces (as it is common practice in large-scale enterprises) are not sufficient.

Within the framework of the project the product genesisWorld was established as the control center for the CRM in the smart organization of the Intranet Partners. The customer-information-system offers a broad functionality with adress-, date-, document- and project-management. The employment of the product genesisWorld is based on a modern client-server-architecture on three layers and makes a good integration and scalability possibility. On this technological basis genesisWorld covers all fields of the customer-and information-management, it can be embedded seamless in the Office-world and offers different mobiles accesses. The CRM-groupware builds bridges among software-applications of each member of the virtual organization and the integration specific to ERP-systems as well as other applications. With the implemented ERP connection an employee of the virtual organization has the possibility to overview with the help of the genesisWorld address history offers, order confirmations, delivery notes, invoices as well as open tasks. In this way one can reconstruct from every workplace of the virtual organization all steps from the offer issue and the order processing to the invoice issue. It is important for the virtual organization that data is know kept up to date by means of automatic data replication between the distributed servers (e.g. address, tasks, ERP data. Thus redundant data management is prevented. Product- and price-catalogues can weave furthermore information texts and pictures directly form genesisWorld in the internet- and intranet pages of each company or virtual enterprise. In the case of a customer approach the Personal Information Assistant (PIA) gives in a self-acting and personalized way information to the network-employee, wherever he may work in the virtual enterprise. The active PIA executes the survey control of customer- and project-histories and informs immediately the user of every change. If for example customers had contact to German network partners or British co-operation partners, the system records clearly this information. The Smart Replication provides for the virtual enterprise or the mobile employee the opportunity to align data of notebooks or among each network partner. Thus the companies can make exactly those data available to their network partner, they want to select and align individually. In this way mobile sales representatives have immediate access to the HTML- or ActiveX-Client of genesisWorld with the aid of a pc or notebook connected to the internet. In so doing the familiar genesisWorld-workplace is available at any time. Further interfaces with restricted browser-functionalities of a lean customer are available for almost every Personal Digital Assistant (PDA).

6. CONCLUSION

The Infranet Project as a best practice case showed how it is possible to realize virtual organizational structures in small businesses. With the combination of groupware and CRM one stood abreast of small division of work and standardization. The IT-based requirements concerning interoperability among heterogeneous IT-architectures of each network partner and the integration of existing old systems in small and medium-sized companies (e. g. enterprise resource planning (ERP)) were also met satisfactory. With the possibility of intelligent replications each enterprise was enabled to feed in non-critical data and information. In so doing one could accommodate the increasing mobility in the economic life. An access is possible at any time via the internet or mobile data communication, as well as via intelligent PDA or notebook-replications. In this way the Infranet Partners achieve their aim, which is the in-network loyalty and customer loyalty according to the quotation of Henry Ford:

- To meet means a beginning.
- To co-operate means a progress
- To stay together means success.

7. REFERENCES

1. Arnold, O.; Faisst, W., Härtling, M.; Sieber, P.: Virtuelle Unternehmen als Unternehmenstyp der Zukunft? HMD – Theorie und Praxis der Wirtschaftsinformatik (1995), S. 8 – 23
2. Blumenthal, N.: Virtual organzational structures. Intranetwork analysis. A empirical study. Research Center of Information Technology. Karlsruhe 2001.
3. Endres, Ruth: Strategie und Taktik der Kooperation : Grundlagen der zwischen- und innerbetrieblichen Zusammenarbeit. 2. Überarb. Aufl.. Berlin : Erich Schmidt Verlag, 1991.
4. Faisst, W., Stürken, M.: Daten-, Funktions- und Prozess-Standards für Virtuelle Unternehmen – strategische Überlegungen, Arbeitspapier der Reihe „Informations- und Kommunikationssysteme als Gestaltungselement Virtueller Unternehmen". Nr. 12/1997 (herausgegeben von Ehrenberg, D., Griese, J., Mertens, P.), Bern, Leipzig, Nürnberg 1997.
5. Fleisch, Elgar; Österle, Hubert: Process-oriented Approach to Business Networking. Institute for Information Management at the University St. Gallen, CH-9000 St. Gallen, Switzerland. in eJOV 2 (2000) 2, Electronic journal of organizational virtualness. http://www.virtual-organization.net.
6. Gillessen, A.; Kemmner, G.-A.: „Virtuelle Unternehmen". Heidelberg. Physica-Verlag, 2000
7. Längsfeld, M.; Weiß, P.: Virtual Organizational structures. An empirical study. Research Center of Information Technology, Karlsruhe 2000.
8. Lindner, T.: Der Mittelstand braucht spezielle Lösungen – Computerwoche 38/2001.
9. Mertens, P.; Griese, J.; Ehrenberg, D. (Hrsg): Virtuelle Unternehmen und Informationsverarbeitung. Springer Verlag 1998.
10. Mowshowitz, Abbe: The Switching Principle in Virtual Organization. Department of Computer Science, The City College of New York, USA. abbem@earthlink.net. Proceedings of the 2nd International VoNet – September 1999. http://www.virtual-organization.net.
11. Picot A.; Reichwald, R.: Auflösung der Unternehmung? In: Zeitschrift für Betriebswissenschaft, Gabler Verlag, Wiesbaden, 1994.

12. Pribilla, P.; Reichwald, R.; Goecke, R.: Telekommunikation im Management – Strategien für den globalen Wettbewerb, Stuttgart: Schäffer-Poeschel Verlag, 1996.
13. Proceedings of the VoNet - Workshop, April 27-28, 1998, http://www.virtual-organization.net.
14. Schwarzer, Zerbe, Krcmar: An eclectic framework for understanding new organizational forms. Discussion paper. Arbeitspapier am Lehrstuhl für Wirtschaftsinformatik. Universität Hohenheim. (510H). 70593 Stuttgart, 2000.
15. Sue Balint and Athanassios Kourouklis: The Management of Organizational Core Competencies. In Pascal Sieber and Joachim Griese (Eds.). Organizational Virtualness. Proceedings of the VoNet - Workshop, April 27-28, 1998.
16. Wendy Jansen, Wilchard Steenbakkers and Hans Jägers: Electronic Commerce and Virtual Organizations. In Pascal Sieber and Joachim Griese (Eds.). Organizational Virtualness and Electronic Commerce. Proceedings of the 2 nd International VoNet - Workshop, September 23-24, 1999.

---

[i] In our paper we use the term smart organization as synonym for virtual organization.

THE FIR E-BUSINESS ENGINEERING INTEGRATION MODEL

Marc Beyer[1], André Quadt[2], Stefan Bleck[3]
Research Institute for Operations Management (FIR)
at Aachen University of Technology (RWTH Aachen)
- E-Business Engineering Department -
e-mail[1]: Marc.Beyer@fir.rwth-aachen.de
e-mail[2]: Andre.Quadt@fir.rwth-aachen.de
e-mail[3]: Stefan.Bleck@fir.rwth-aachen.de

In the recent years, the broader application of web-based technologies caused a rapid development within the entrepreneurial field. In order to exploit first-mover advantages, enterprises often preferred a quick-paced introduction of E-Business solutions, hence neglecting more holistic and integration-related issues. Therefore, such E-Business solutions were usually simply and hastily embedded into the existing business processes and organisational structures. As a result, E-Business projects often did not reach the striven targets or even failed, with the consequently growing lack of trust towards the above-mentioned business approach.

E-Business, Practical Application, Holistic Model, Integration, Study Results

1. HIGH COMPLEXITY AND HIGH POTENTIALS

Current E-Business projects tend to be more and more complex. As a result, managing and realising them successfully becomes more difficult. On the one hand, complexity of E-Business solutions is growing rapidly, on the other hand, their introduction usually concerns several departments within an enterprise. Therefore, companies need practice orientated methods, to fulfil the necessary tasks in a clearly structured way. To develop such methods and to point out needs for action is one of the main attentions of the *Research Institute for Operations Management (FIR)*.

Several field studies underline that the potentials of both stationery and mobile Information and Communication Technologies (ICTs) are nowadays exploited insufficiently. On the other hand, it is remarkable that the fast exchange of information through high performance digital networks perfectly suits the striven achievement of a higher efficiency and transparency within a company and thus improved productivity (Meitner, 1996). In order to be successful in their E-Business implementation, enterprises have to make use of a methodical and structured approach that enables them to carry out customer demands more efficiently (Römer, 1997; Herrmann, 1999). In particular, this process must e.g. involve a systematic analysis of all possible applications of different E-Business solutions regarding the fundamental ICT potentials. Furthermore, the existing company resources (e.g. the current product and service portfolio) as well as the involved business processes and organisational structures have to be accurately taken into consideration.

Within our daily contact with the industrial world, we have noticed that *enterprises currently lack of such methods and structured procedures.* To analyse this phenomenon more precisely, the FIR E-Business Engineering Group conducted a study which investigated the obstacles as well as the potentials and underlying methods of E-Business projects in more than one hundred enterprises.

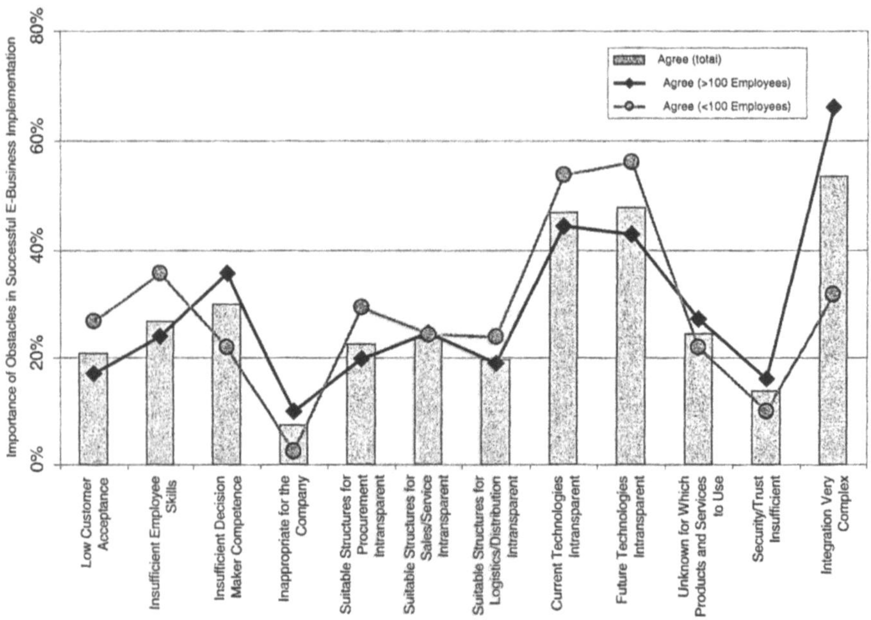

Figure 1 – Especially small and medium sized enterprises lack structured E-Business approaches (Bleck et al., 2001)

The results of this study at the same time demonstrate the high potentials, which characterise E-Business solutions. As shown in figure 2 even small Enterprises could benefit from these potentials in some fields. Furthermore other sources also hint at the potential for increased productivity (Dhillon et al., 1996; Apostolopoulos et al., 1996). Despite the current depression in several sectors of the so-called new economy and the shyness of many companies regarding the introduction of internet technologies, the survey clearly showed, that the utility of E-Business projects is high, if conducted systematically and well founded.

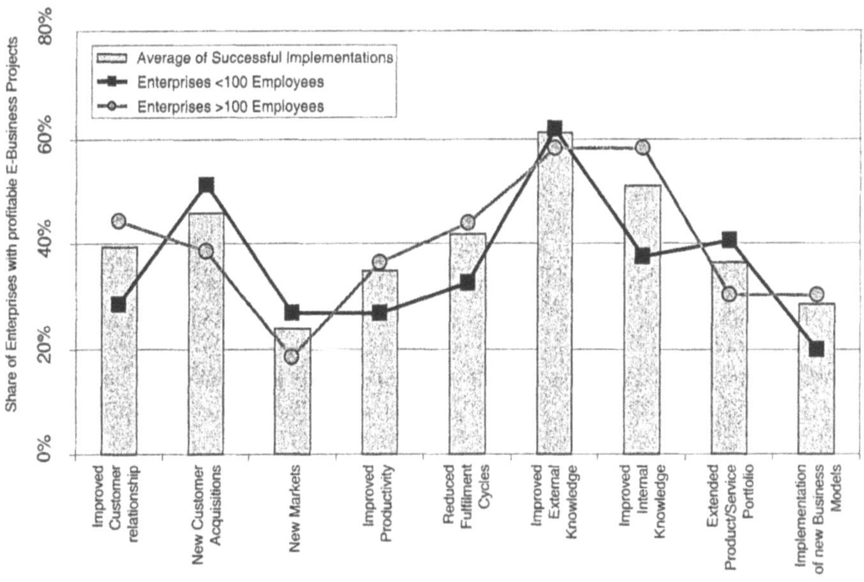

Figure 2 – Even small Companies can gain strong advantages from e-business projects (Bleck et al., 2001)

2. THE FIR PHASES-MODEL OF SUCCESSFUL BUSINESS ENGINEERING

In order to ensure a structured and methodical approach with the introduction and further development of E-Business, the FIR pursuits a phase-based model. As depicted in figure 3, four different phases are conducted, which are each supported by specific methods systematically (see figure 4).

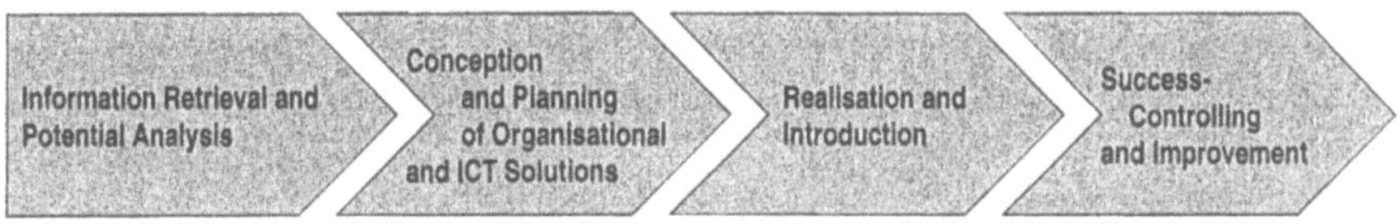

Figure 3 - Phases-Model of successful Business Engineering

Phase 1: Information Retrieval and Potential Analysis
In this phase the company has to develop a suitable enterprise-specific E-Business strategy. Therefore several methods for strategy development and definition can be used. Particularly helpful are market overviews or technology monitoring or screening. Furthermore creative techniques have to be applied, in order to derive promising solutions from technological potentials. Additionally the FIR has developed the "Innovation Workshop", a single day training and discussion event

that helps companies to identify potentials, define their strategic goals and find suitable E-Business policies.

Phase 2: Conception and Planning of Organisational and ICT Solutions

This phase makes use of reference models and example implementations. Additionally classical methods of business reengineering and cost/benefit analysis are applied. Economic success is verified by a first profitability analysis based on experts' estimations. In order to ensure an early consideration of integration aspects, according approaches are implemented.

Phase 3: Realisation and Introduction

The particular realisation and introduction of E-Business solutions is supported by methods and tools of organisational development. These methods are to be completed in respect to information technological integration and the design of interfaces to existing systems. The optimal dates for introduction are determined e.g. by the use of a technology calendar or road maps.

Phase 4: Success Controlling and Improvement

Finally a systematic assessment of the implemented changes should be conducted by the means of predefined criteria. This can be achieved internally by a detailed cost/benefit analysis or externally by a benchmarking process.

The methods used in the different phases depend on the particular project respectively problem. Figure 4 shows, what methods are useful in which phases.

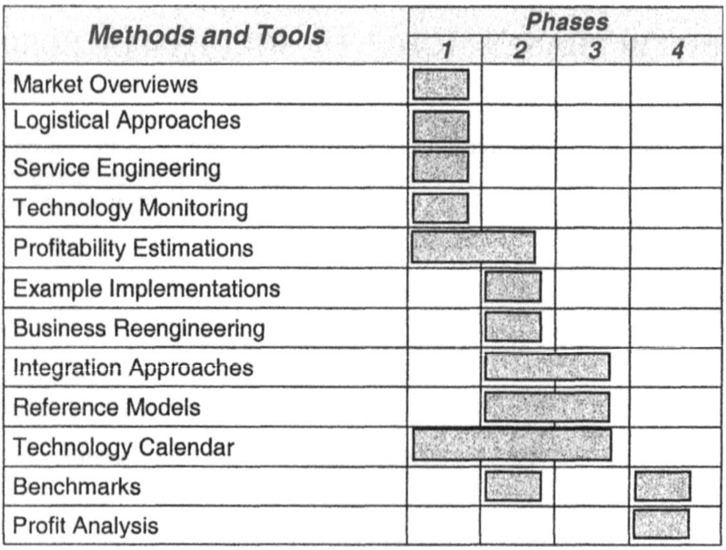

Figure 4 - Applications of Methods in the Project Phases

The application of these methods and procedures can not be considered separately. On the one hand they affect every organisational level in the company. On the other hand they influence each other and thus have to be regarded as a whole.

The choice and introduction of suitable methods and E-Business solutions is a complex problem. Experiences from similar projects, e.g. from reorganisations, have shown, that a holistic approach is necessary and more successful than focussing on certain parts or the problem.

It is therefore consequential to abstract the actual methods and procedures into an integration schema. Thereby it is possible to show and manage the relationships between the result of strategy planning, operational implementation and success control.

Moreover E-Business solutions do not only have to be developed, but also operated and maintained. This demands a methodically founded planning tool. It also helps to identify E-Business components, that can be reused or applied in multiple fields.

These interactions and interdependencies of the different methods of E-Business as well as the complexity of the aligned problems is regarded by the FIR in a special "integration model", that is explained in the following section. It's different planes are the design areas and relevant fields of action that define the structure for the development and application of suitable methods.

3. THE FIR E-BUSINESS INTEGRATION MODEL

E-Business involves all entrepreneurial activities in which the application of ICTs plays a decisive role. It integrates approved methods of Business Engineering as well as the potentials of the new media and technology. Within this context, we define *E-Business Engineering* as "all the methods and procedures that support companies of different industrial sectors to systematically develop, implement and run E-Business solutions" (Forzi et al., 2003). E-Business Engineering can be hence defined as the *systematic design and implementation of E-Business solutions and models*. It also contains the aspects of Clausing (Andrade et al., 1995) and Clausing and Witter (Clausing et al., 1992).

The FIR E-Business Engineering Research Group focuses on the development of both *inter-organisational* (Business-to-Business) and *intra-organisational* (within the departments of a single company) solutions for small and medium sized enterprises (SMEs) of the manufacturing industry, e.g. for procurement, sales and services. This includes explicitly analysis, modelling and design of E-Business solutions for new intermediators, i.e. market participants who enable transactions between suppliers and demanders (Baily, 1998).

As previously underlined, the development of ICT based solutions needs a systematic design. Our research institute developed a meta-method to support enterprises by the company-wide integration and implementation of E-Business solutions. Such a holistic approach exploits several established methods from the fields of product and service engineering, business organisation and computer science (e.g. methods of business modelling, service engineering, process and system design, technology monitoring and planning).

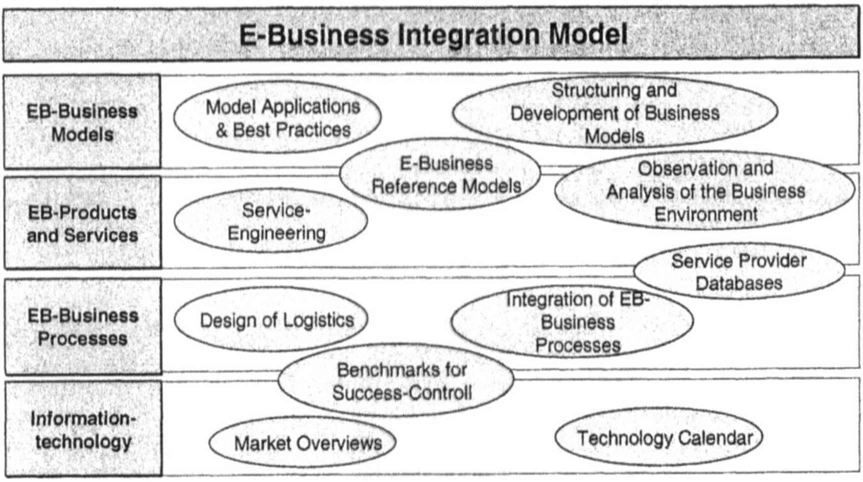

Figure 5 – The E-Business Integration Model Defines Four Areas of Research and Development

The above-mentioned methods as well as their interactions represent the core elements of our *E-Business Engineering Integration model.* Our model focuses on the following fields:

E-Business ***business models*** as the upmost plane represents the design of E-Business architectures for an efficient inter-organisational exchange of products and services within manufacturing networks. The main focus is on benefit and revenue relationships.

E-Business ***products and services*** means systematic development of E-Business product and service portfolios for manufacturers and service providers.

E-Business ***Business Processes*** concerns the analysis of the entrepreneurial changes triggered by E-Business, especially in respect to business activities and processes that are heavily supported by electronic information and communication technologies

Information and Communication ***Technologies*** are the basis for all E-Business activities. Analysis of existing and prospective ICTs, as well as methods of technology management and planning are the basis for their effective use.

The objective of our approach is therefore to discuss the different levels and methods of our integration model, as well as its operational value.

4. SYNOPSIS

The results of our study demonstrate the high potentials of modern information and communication technologies. At transforming these potentials into specific and efficient solutions a systematic approach is inevitable.

Nonetheless E-Business centred methods and approaches for the conception and realisation of E-Business are not widespread yet. To close this gap is a main focus of the FIR. The resource orientated integrated holistic model that we developed is our basis to achieve this goal. The model allows the construction of a toolbox of methods by the application and evaluation of classic approaches as well as the integration of newly developed methods.

Regarding the results of our study, that demonstrate the difficulties with integrating innovative E-Business solution in conventional processes, the methods of business reengineering and organisational analysis seem suitable, yet insufficient. They need to be completed by newly developed procedures, that support companies to take advantage of the technological potentials in modern ICT.

5. REFERENCES

Andrade, R. Clausing D. P. Strategic Information of Product, Technologies and Core Competencies with support of Planned Reusabilitiy. Draft internal working paper at MIT, Boston: 1995.

Apostolopoulos, T. K., Pramataris, K. C. Information Technology Investment Evaluation: Investments in Telecommunication Infrastructure. International Journal of Information Management, 16(1996)4; 287-296.

Baily, J.P. Intermediation and Electronic Markets: Aggregation and Pricing in Internet Commerce. MIT. Boston: 1998.

Bleck, S. Methodeneinsatz bei der Erschließung des E-Business – Ergebnisse einer Studie. Industrie Management, (2001).

Clausing, D. P.; Witter, J. H. Integration of Reusability and Interface Management into Enhanced Quality Function Deployment Methods. Working Paper, 1992.

Dhillon, G., Backhouse, J. Risks in the Use of Information Technology Within Organizations. International Journal of Information Management, 16(1996)1; 65-74.

Forzi, T. and Laing, P. E-Business Modeling. In Albalooshi, F. (Edt.) Virtual Education: Cases in Learning & Teaching Technologies. Hershey, London, Melbourne & Singapore: IRM Press, 2003; 113-138.

Herrmann, G. Sichere Geschäftstransaktionen. In Turowski, K. (Edt.) Tagungsband des 1. Workshops Komponentenorietierte betriebliche Anwendungssysteme (WBKA1). Magdeburg: 1999; 89-93.

Meitner, H. Infrastruktur für die Geschäftsprozeßoptimierung – die Unterstützung durch Infobahnen. FB/IE Zeitschrift für Unternehmensentwicklung und Industrial Engineering, 45(1996)1; 16- 20.

Österle, H. Business Engineering Prozeß- und Systementwicklung. Berlin et al.: Springer, 1995.

Römer, M. Strategisches IT-Management in internationalen Unternehmungen. Wiesbaden: Gabler, 1997.

MICROPLANO – A SCHEDULING SUPPORT SYSTEM FOR THE PLASTIC INJECTION INDUSTRY

Cristóvão Silva
Mechanical Engineering Department, Coimbra University, Portugal.
cristovao@gestao.dem.uc.pt

Luís M. Ferreira
Portuguese Catholic University, Figueira da Foz, Portugal luis.ferreira@crb.ucp.pt

In this paper we introduce a production-scheduling problem found in a Portuguese small medium enterprise (SME) that produces small plastic parts for the electric/electronic industry. The parts are obtained by plastic injection onto a mould, in a shop composed by 32 parallel machines. The characteristics of the production system are presented and a two-step algorithm, developed to schedule all the jobs, is referred. This algorithm has been incorporated in a decision support system (Microplano), which also contains a database management module and a user interface module. Microplano has been implemented in the company and is being in use on a regular basis.

Keywords: Decision support system, Lot size scheduling, Plastic injection industry

1. INTRODUCTION

This case study was conducted for a Portuguese SME that produces small plastic parts, essentially for the electric/electronic industry. The parts are obtained by injecting melted plastic onto a mould. This operation is conducted in a workshop composed by 32 injection machines, in parallel, and not identical. After being injected some parts are finished while others need to suffer some assembly operations made manually. The production planner used to generate plans manually, but due to the rapid grow of the company the managers wanted to improve the production planning process. In this paper a decision support system developed for the above referred company is presented. This system was developed to elaborate production plans for the injection department of the company where two objectives were desired:

- Respect the due dates, or at least minimize any delays.
- Minimize the number of mould changeovers.

The decision support system has been implemented in the company, where it has been in daily use.

The paper is organized as follow. In section 2 the company scenario is described in detail. Section 3 presents the algorithms choose to elaborate production plans. In section 4 Microplano, the Decision Support System that incorporate the algorithms, is presented.

2. THE COMPANY SCENARIO

This company produces small plastic parts for the electric/electronic industry, selling them to international customers. Two times per year the major costumers send to the company a six months reservation, indicating a forecast of their needs for this period. This information is updated monthly with the probable requirements for the next month. Finally bi-weekly firm orders are received, indicating precise quantities and due dates. The company manufacture a wide range of product, injected in more then 600 different moulds, owned by the clients, but managed by the company. To satisfy the customer orders the company must plan the execution of more then 300 production orders per week.

The required products are manufactured by plastic injection onto a mould composed by two metallic pieces. Once these two pieces are closed together the mould forms a cavity with the shape required for the final product. In each cycle the machine close the mould, inject melted plastic to the cavity, open the mould and eject the plastic parts. After the injection operation some parts are finished while others need to suffer some finishing operations. These operations are conducted in the assembly department of the company and mainly consist in:

- Assembly two or more parts to obtain a finished product.
- Add some metallic components to the plastic parts.

A mould can have one or more cavities, and these cavities can have the same or different shapes. Therefore in each cycle it is possible to obtain one part, various identical parts or various different parts. Various parts produced in the same or in different moulds may compose a finished product.

The company production system consists of a shop composed by 32 machines in parallel. The machines are not identical and therefore a mould can only be affected to a small group of machines. The machines differ in the pressure they can exercise and in the ability to automatically introduce small metal components into the parts. The scheduling problem arises in the injection department where the scheduler must decide in which machine shall each required mould be mounted, and the processing sequences of all jobs affected to the same machine.

Before the implementation of the system the production planner generated the schedules manually. The resulting schedules were introduced manually onto Microsoft Project to generate a Gantt chart to send to the operators of the machines.

3. THE ALGORITHMS

3.1 Literature Review

The problem described in the previous section is known as the lot size scheduling problem. In this problem the jobs can be grouped by type, where jobs of the same type don't require a machine changeover between them. Changing from production of one type of job to another requires a change of machine instalment to another state. As these changeovers are time consuming the objective is to minimize the number of changeovers, while respecting the jobs due date. In the problem under analysis jobs of the same type are those who can be produced in the same mould.

The lot size scheduling problem has been addressed by various authors, and many models, involving different features, can be found in the literature. The problem can consider single or multiple machines, and in the later case the machines can be parallel or in sequence. The lot size scheduling problem formulation can also involve setup costs that can be fixed, product dependent or sequence dependent. Another feature is the division of the planning horizon in small or large buckets. For a survey on lot size scheduling problem models, see for instance Drexl and kimms (1997).

The lot size scheduling problem has been addressed by various authors for the single machine case, considering the planning horizon divided in identical small bucket, and no backlogging allowed, see for example Glassey (1968), Gascon and Leachman (1988) or Blocher and Chand (1996). In Blazewicz et al (1996) a polynomial algorithm for two jobs is presented which solves optimally the problem for the single machine case. In the same book the algorithm is generalized to solve the multiple identical machine case. Pattloch et al (2001) develops and evaluate a heuristic, based on the models proposed in Blazewicz *et al* (1996), using it in an industrial case.

Meyer (2002) proposes an algorithm to solve a more generic lot size and scheduling problem, combining the local search metastrategies threshold accepting and simulated annealing, respectively, with dual reoptimization. In Santos-Meza *et al* (2002) a lot size and scheduling problem found in a foundry, with particularities similar to those found in our case study company, is presented. In their paper these authors conclude that general heuristics, based on branch and bound method, will scarcely be better than problem-specific heuristic procedures to obtain a feasible solution within an acceptable running time.

For a survey of different methodologies proposed by authors to solve the lot size and scheduling problem, see for instance Staggemeier and Clark (2001).

From our literature review, and considering the specificities of our problem, we decide to develop a simple methodology to solve it. We believe that this methodology, correctly supported by a decision support system, could help the production planner of the company to obtain adequate production schedules to the injection department.

The methodology we propose is based in two heuristics presented in Sule (1997), described in section 3.3 and 3.4 of this paper. Those heuristics, that were not intended to solve the lot size scheduling problem for non identical parallel machine, can, has we will see; provide a solution for the problem when used sequentially.

3.2 Scheduling preliminaries

The orders are not sequenced as they are received from the customers; instead some previous operations are required to generate jobs and their characteristics. Those operations are required due to the stock level of finished products and components, parts injected in the same mould and orders for the same product and due date, and are described below.

In a first step quantities existing in the final product inventory are discounted from the orders and a list of all finished product required is generated. The resulting needs for finished products are exploded to obtain a list containing the components requirement. From this second list the quantities existing in the components stock must be discounted. The result of these operations is a third list containing all the parts that need to be injected to satisfy the costumers order. In a next step this list is checked to find production orders that are produced simultaneously in the same mould. The quantities of the different orders must be adjusted to take this fact into account. For example suppose two production orders produced in the same mould: one for 200 X and one for 50 Y. The mould where X and Y are produced have two cavities to produce X and one cavity to produce Y. In this case the second production order must be adjusted to 100 Y. 50 of these components will be used to satisfy the costumers' orders while the others will go to stock.

The production orders of products required to the same due date and produced in the same mould are grouped into one job. Therefore a job can satisfy one or more production orders. The final step required prior the sequencing of jobs consist in determining the processing time of each job. This time is a function of the mould cycle time, number of cavities and the number of parts required.

Once all jobs had been generated they must be sequenced in the injection department. In order to simplify this problem we decide to obtain the schedules in two steps:

- In the first step the moulds are affected to the machines, with the objective of distribute evenly the workload among the different machines.
- In the second step the jobs imposed by the moulds affected to a machine are sequenced, in order to minimize the delays and the number of changeovers.

To accomplish these two steps we used two heuristics presented in Sule (1997) and described bellow.

3.3 Heuristic 1 – Affecting jobs to the machines

We consider that once a mould is affected to a machine, all the jobs processed by this mould are affected to that machine. Therefore each mould will impose a load to the machine where it is affected equal to the sum of the processing time of all jobs produced in this mould. Given this assumption a four step algorithm, described below, was choose to affect each mould to a machine.

Step 1: Calculate the flexibility index for each mould (the number of machines on which the mould can be processed) and for each machine (the number of mould that can be used in the machine).

Step 2: Select the current least flexible mould, if there is a tie brake choose the mould associated with the least flexible machine; assign the mould to the machine with the least workload.

Step 3: Recalculate the machine workload, considering the load imposed by the affected mould.

Step 4: Modify the flexibility index of all the machines on which this mould could have been mounted. Repeat step 2, 3 and 4 until all the moulds are assigned to a machine.

3.4 Heuristic 2 – Sequencing jobs

At this stage we have, for each machine, a list of jobs that must be processed by it. Each one of these jobs has a due date and a processing time associated, and must be sequenced. To accomplish this objective we choose the heuristic presented below.

Step 1: Group the jobs by job types – jobs processed in the same mould - and within each group arrange the jobs in the ascending order of due dates.

Step 2: Calculate the maximum start time (MST) for each job. MST is defined as Due date – (processing time + setup time + assembly time).

Step 3: Start the job sequence with a job that has the minimum value of MST and calculate the sequence time (i.e., time associated with the sequence so far).

Step 4: Continue the sequence by assigning the next sequential job from the same group until either all jobs are assigned or the sequence time exceeds MST of an unsigned job in some other group. Denote this job by K. Go to step 5.

Step 5: If all job are assigned stop. If not, remove the last job that had caused the sequence time to exceed the MST of K and assign job K in the sequence instead. Calculate the new value of the sequence time and go back to step 4.

These two algorithms and the operations required prior the scheduling operation were implemented in a decision support system described in the next section.

4. THE DECISION SUPPORT SYSTEM

Each time the scheduler wishes to make a production plan, all customer orders are downloaded from the company information system. The downloaded file consists of all customer orders entered by the marketing department since the last production plan. These orders are added to the Microplano database generating a file of all customer orders received by the company and that have not yet completed production.

4.1 Sequencing preliminaries

The first step realized by Microplano consist in analyse the file containing customer orders and generating the jobs and their characteristics as described in §3.1. At this stage Microplano presents, in a set of screens, the relevant information to the user: mould and machine availability (they can be in maintenance), the matrix mould/machine, and the production orders required to fulfil the procurements. After this the planer is conducted through two screens. In the first screen a list of the production orders required is presented, and the user can delete or add new production orders. These capabilities has been used to eliminate orders cancelled by customers or create new orders when there is low loads of work and the planer decide to make stocks of some products. The user can also access the matrix mould/machine and change it. In this screen the user can also change the machine and mould availability, indicating periods where they will be unavailable, due to maintenance operations. When all jobs have been generated the DSS go to the next step "loading the machines".

4.2 Loading the machines

At this stage Microplano affect each mould to a machine, using the heuristic presented in § 3.2. The results are presented in the form of a table where the load imposed to each machine is presented and in the form of a graph (see Figure 1).

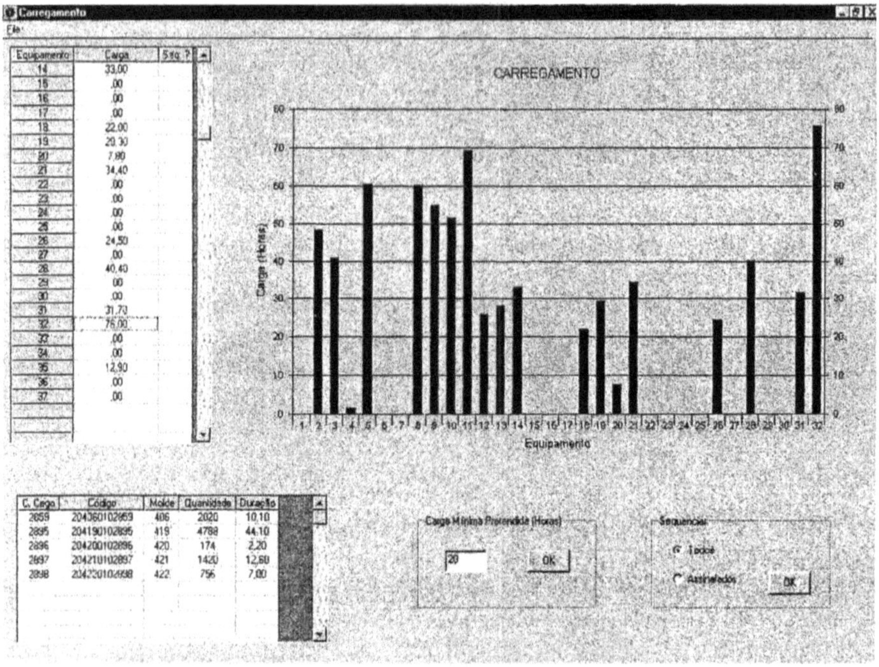

Figure 1 – Screen presenting the workload imposed to each machine

During the implementation of the system the scheduler found that it was usual to have great differences between the loads of the different machines. This fact was due to the differences of work imposed by the different moulds. We decide to allow the user to impose a minimum load to the machines. During the duration of the generated plan the machines with workloads under this minimum will not work, and the jobs affected to it are transferred to other machines. This information can be used to plan the maintenance work.

4.3 Sequencing work

The sequences obtained are presented, for each machine in a sheet. The jobs that will be completed with delays are presented in red, and when a changeover is required between two jobs it code is presented in bold. The number of late jobs and changeovers are presented in two text boxes, to allow the user to be quickly aware of the values of performances measures. The screen where the sequence is presented to the user is shown in Figure 2.

The user can act upon the schedules proposed by Microplano, swapping the position of one or more jobs. The planner has essentially used this possibility to look for reduction in the number of changeovers, moving jobs to a position next to a job made in the same mould. This swapping generally conduces to a sequence with less changeovers but more late jobs. The user can know the effect of the changes just looking to the text boxes indicating the performances measures.

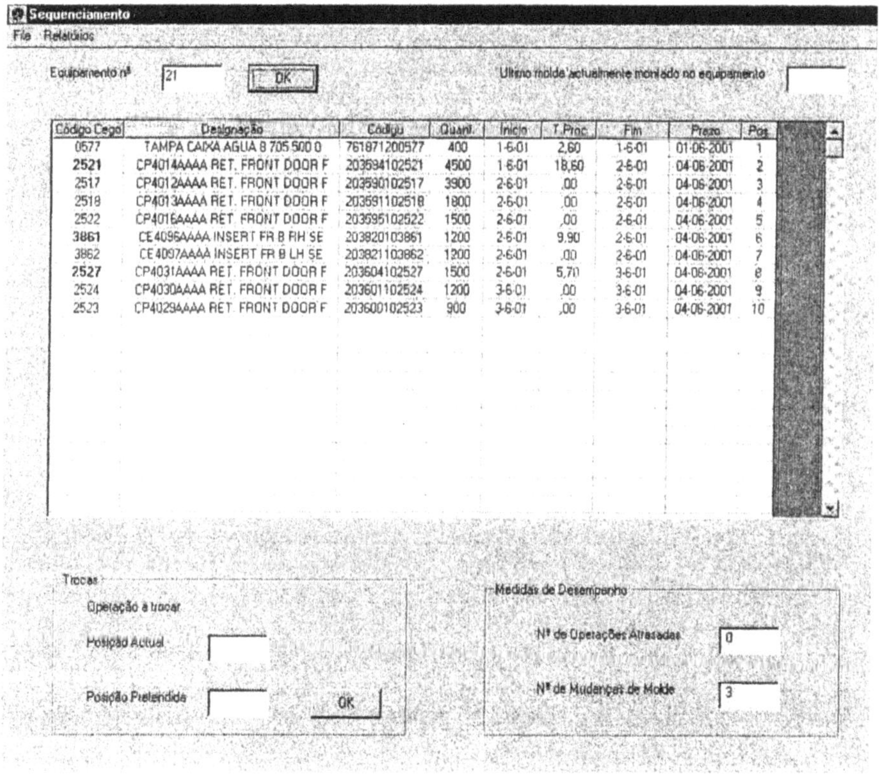

Figure 2 – Screen presenting the processing sequence for a given machine

4.4 Generating reports

When the scheduler considers that he obtained a satisfactory plan, he can use Microplano to generate some reports. The first report is a sheet containing the obtained production plan; where for each machine a list of all jobs are presented. For each job the system gives the start date, the processing time, the finish date and the due date. This information can be automatically transferred to Microsoft Project who generates the Gantt chart to send to the operators.

Microplano also generates reports containing information about the requirements for raw materials, components and packages to satisfy the generated plan. This information is send to the purchasing department which can use it to define corrects acquisition policies.

4.5 Operating Microplano

Microplano can be operated in two different ways:

- Considering that there is no job currently affected to the machines, i.e., ignoring the previous obtained production plan.
- Considering the previous obtained plan, i.e., considering jobs which are actually being performed in the machines.

Let us begin with the second way, the one used more often and more adequate to generate weekly plans. In this case Microplano access Microsoft Project, where the previous plan is being updated, and analyse the current state of jobs sequenced in the previous week. The jobs currently being processed in each machine, and the second job of the sequences obtained in the previous plan are fixed. All other jobs return to the Microplano database. Microplano calculates the time required to complete the jobs previously fixed and sum it to the current time given start time for each machine in the new plan.

This procedure will have two main implications in the procedure followed to establish the required production plan:

- All the moulds required to process fixed jobs from previous plans are kept in the machine where they were affected. Consequently, all jobs processed in these moulds are also automatically affected to these machines.
- When the sequence is presented to the user, a text box containing the code of the mould actually in operation in a machine is shown to the user. This will allow trying to swap jobs in order to put in the first place of the new sequence a job produced in that mould, seeking a reduction in the number of changeovers.

The other way to operate Microplano, ignores what is actually happening in the shop, considering that all machines are idle. This way is simpler and gives less work to the planner, but is not adequate to establish weekly plans. This procedure is being used to generate long terms plans. In this case Microplano considers not only firm costumer orders, but also the reservation make by them, which can or not become firm orders. This procedure is used to have a forecast of the loads that will be imposed in the next few months identifying possible capacity problems.

5. CONCLUSIONS

The system has been implemented successfully in the factory and has been in daily use now for one year. The production planner is following the generated schedules. The quality of the schedules appears to be significantly better. The reason lies in the fact that the scheduler now dedicates most of his time to the manipulation of schedules, while receiving immediate feedback on how performance measures are being affected.

We intend to develop further Microplano, the decision support system presented in this paper. These developments will be essentially the introduction of different algorithms, seeking if better sequences can be obtained. The algorithms we intend to implement and test are basically of three types:

- An algorithm to affect and sequence jobs simultaneously, and not in two steps like the ones presented in this paper.

- An algorithm where jobs produced in the same mould can be affected to different machines.
- Metaheuristics to automatically swap jobs positions to find improvements in the obtained sequences.

6. REFERENCES

Blazewicz J, Ecker KH, Pesch E, Schmidt GS and Weglarz J. "Scheduling Computer and Manufacturing Processes", Springer-Verlag, Berlin, 1996.

Blocher JD and Chand S. "A Forward Branch-and-Search Algorithm and Forecast Horizon Results for the Changeover Scheduling Problem", European Journal of Operational Research, 1996, 91: 456-470.

Drexl A and Kimms A. "Lot sizing and scheduling – survey and extensions", European Journal of Operational Research, 1997, 99: 221-235.

Gascon A and Leachman RC. "A Dynamic Programming Solution to the Dynamic, Multi-Item, Single-Machine Scheduling Problem", Operations Research, 1988; 1: 50-56.

Glassey CR. "Minimum Change-Over Scheduling of Several Products on one Machine", Operations Research, 1968; 342-352.

Meyer H. "Simultaneous lot sizing and scheduling on parallel machines", European Journal of Operational Research, 2002, 139: 277-292.

Santos-Meza E, Santos M O and Arenales M N. "A lot-sizing problem in an automated foundry", European Journal of Operational Research, 2002, 139: 490-500.

Staggemeir A T and Clark A R. "A survey of lot sizing and scheduling models", 23rd Annual Symposium of the Brazilian operational Research Society (SOBRAPO), Brazil, 2001.

Sule D. R., Industrial Scheduling, PWS Publishing Company, Boston, 1997.

Pattloch M, Schmidt GS and Kovalyov MY. "Heuristic Algorithms for Lotsize Scheduling with Application in the Tobacco Industry", Computers & Industrial Engineering, 2001; 31: 235-253.

QUALITY CERTIFICATION IN THE VIRTUAL ENTERPRISE

Ângelo Martins, J. J. Pinto Ferreira, J.M. Mendonça
ISEP (amartins@dei.isep.ipp.pt)
FEUP-DEEC / INESC Porto
FEUP-DEEC /Fundação Ilídio Pinho

Within the scope of virtual and networked enterprises, this paper introduces a novel methodology for the integration of quality management into enterprise modeling and allowing for quality certification of the whole enterprise. The ISO 9001:2000 standard is presented as an interesting starting point for the virtual enterprise quality management system implementation. Quality management system certification within the scope of ISO 10011 (or even ISO FDIS 19011) is however quite a difficult task. An alternative scenario for certification is therefore proposed.

Keywords: Quality Certification, Virtual enterprise

1. INTRODUCTION

For a company to control adequately the products or services (or both) it offers to the customers, it has to achieve global control of its supply chain. This is not an easy task, and only with heavy client bargain power can this be successfully achieved. That is the case of the automotive industry, where manufacturers imposed systems for information exchange (mainly EDI-Electronic Data Interchange) and a universal quality control system (QS-9000 and ISO/TS 16949) on their suppliers. And, even for companies with a vast experience in quality management across vast supply chains, as it is the case of Ford Motor Company, things can often go wrong. As an example, the new 2002 Explorer has been recalled twice in its first three months on sale and the Escapade SUV was recalled five times in its first year (BusinessWeek, 2001). Quality assurance in the large and complex supply chains that build today's virtual enterprises is definitely an area where considerable organisational improvement is to be expected.

An example of a structure where the organisation is built around the supply chain is the Virtual Enterprise (VE), whose basic structure is depicted in figure 1. According to Camarinha-Matos, "A virtual enterprise is a temporary alliance of enterprises that come together to share skills or core competencies and resources in order to better respond to business opportunities, and whose co-operation is supported by computer networks" (Camarinha-Matos, 1999). In such a collaborative environment, the upper management's role becomes that of a facilitator and its authority is somewhat diluted. On the other hand, the VE has no assets of its own, using instead members' assets. Each member decides which assets can be used by the VE and when, these becoming the virtual assets of the VE. A contract of some kind (physical, electronic, etc.) is required as to define the member's rights and obligations related to the VE and also the VE operation rules.

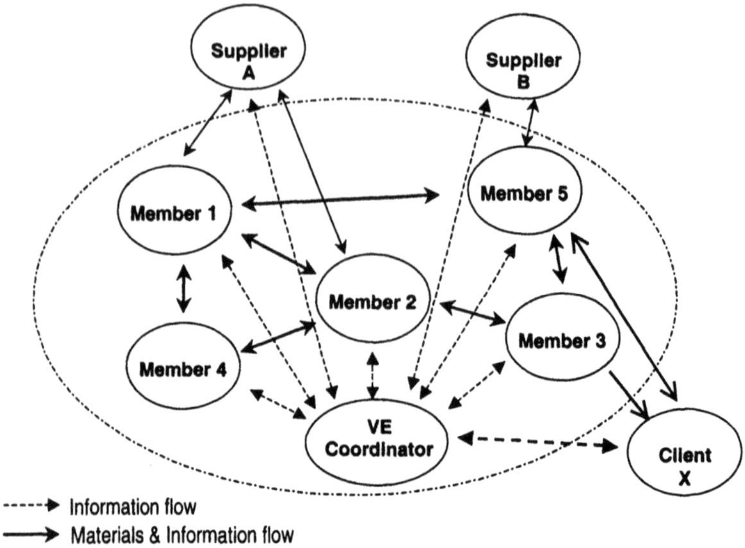

Figure 1 – Example of a VE structure

Another important characteristic of the VE environment is its continuous change (figure 2). In the creation phase, a member structure is assembled, together with the definition of member rights and obligations. Though this also happens in any enterprise, in a VE, partner search and selection is not limited to the creation phase: it may actually happen frequently during its lifetime, with extensive changes to the supply chain size and geometry. Such changes may result from unexpected events, such as a VE member failing to fulfil its obligations, thus having to be replaced; or even result from the natural evolution of the VE. Actually, partner selection may be even seen as a special case of general procurement where the supplier requirements are more stringent and the VE structure will itself change according to supplier's characteristics.

The seamless integration of several enterprises to work together as a VE is an extremely complicated engineering project. To help on the organisation of dispersed knowledge and serve as a guide in enterprise integration programs, several reference architectures have been presented. These reference architectures don't present cookbook like recipes, but define instead areas of knowledge, requirements, IT tools, workflows, etc. One of such reference architectures is the Purdue Enterprise Reference Architecture (PERA) (Williams, 1993).

The combined use of reference architectures and models allow for an easier implementation of the enterprise project. The model conversion to the real system has always been a delicate process, with many semantic problems and associated errors. To overcome this, the model enactment concept was developed (Pinto Ferreira, 1999). Instead of translating the model to an implementation language, the idea is to present an IT infrastructure to support model execution. This not only avoids mistakes, but also allows for a faster implementation. Nevertheless, the chance for error is always present and being the VE so reliant on IT technology, both VE's engineering tolls and operating system have to be addressed in a VE quality management system.

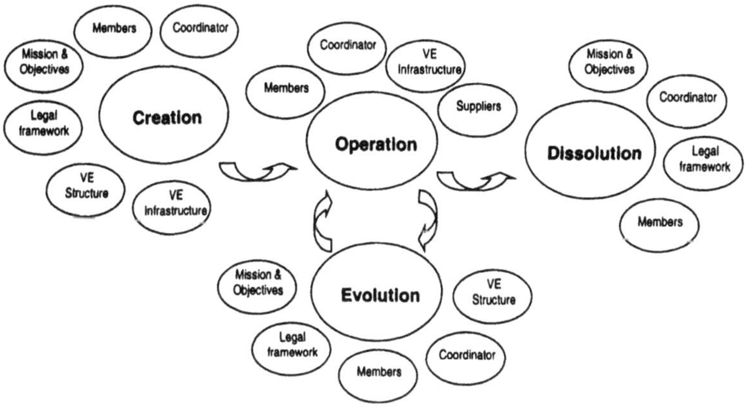

Figure 2 - VE life cycle (Martins, 2000)

2. QUALITY AND THE VE

Never before there was such awareness about quality management as there is today, though many times there is not a clear distinction in people's minds between quality and the ISO 9000 set of standards. In fact, ISO 9000 worldwide popularity make its use almost mandatory when developing a Quality Management System (QMS) for the VE, though we must recognise that some improvements/adaptations should be seriously considered.

In the VE world, quality, and especially QMS certification, could be an interesting way to remove some of the barriers that hinder VE expansion as an alternative enterprise organisation structure. One of these is the natural lack of trust in a type of organisation with no physical assets of its own, with a virtual head office in a tax heaven, and whose "motto" is actually to be "short-lived". An example of VE could be the association of a group of design centres and manufacturing plants, placed around the globe so that they can offer around the clock design services and convenient manufacturing in every major market. Each member may still operate alone in its own market, but through the VE, it increases its reach to the global market, without risking its independence. Indeed, VE could be a low risk approach for SMEs' globalisation.

The problem is that, though it may be technically feasible, it is even more complicated to define a common legal ground for the VE that can be accepted in every market. In fact, regulations are still very different from market to market (country or commerce zone), making it very difficult for a simple group of SMEs to have a global reach. In the end, the VE is very much like a global corporation, except that it lacks the huge resources these companies usually have. A low cost quality management system certification process could make a difference indeed.

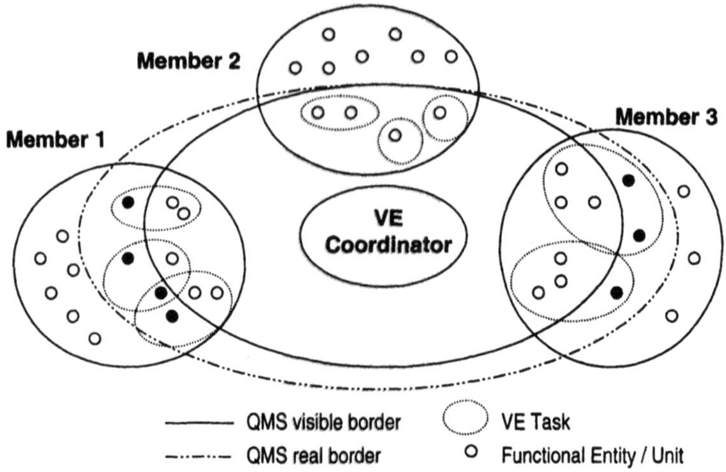

Figure 3 – VE QMS scope

An important step in the definition of a QMS is the system's scope. In the VE, this is not straightforward. In figure 3 a simple example of a VE is presented. Each task (doted circle in the figure) is implemented by member's units (small circles). When assembling the VE, each member allocates tasks and related units to the VE operating system, but these units can't be limited to the ones that directly implement the tasks (visible border in the figure). Units with supporting roles (filled small circles) should also be included, extending VE's QMS border. The problem is how to know if each member is allocating the right units. And we cannot forget that too many units make the system larger and more difficult to manage, not to mention the problem that these units are part of two different QMSs, the member's and the VE's. The solution may lay in using lots of common sense and QMS standards as well.

2.1 ISO 9001:2000 AND THE VE

Perhaps the best way to start the study of ISO 9001:2000 application to the VE is from the top, i.e., the management. While VE operation may be not conceptually different from the operation of a common enterprise one, the same may not be said about VE management. In a common enterprise, there is a clear definition of management responsibilities, these being well understood by both internal and external stakeholders. In the VE management/co-ordination responsibilities will be implementation dependent (Martins, 2000). To comply with the standard, we believe it will be enough to present a clear definition of a management structure with enough authority to implement each of ISO 9001:2000 requirements, either in the VE or in the member enterprises.

Still on the management issues, the ISO 9000 set of standards focus on enterprise evolution through incremental steps (quality improvement, etc.) and the complete enterprise life cycle is not covered (Figure 2). There are several possible explanations for this. In fact, though quality management system formal design may be considered from an early stage when some new enterprises are created, that is not certainly the case for most small and medium enterprises. It is only when the enterprise is fully operational and already working for some time that a QMS is formally designed and implemented. Actually, most certification bodies define a mandatory QMS operation period (a few months) prior to its certification. However, if quality certification is to be of any use in the VE, then it must be considered since its creation. We believe the best way to accomplish this is to include quality requirements in the development of enterprise models and the IT infrastructure. Moreover, enterprise models should describe the four phases of the VE life cycle (We have here the typical chicken and egg problem, but we will elaborate on this later!).

Secondly, common enterprises are not supposed to close, unless perhaps in case of bankruptcy. Being the VE essentially a temporary enterprise, it becomes mandatory to define *post mortem* guarantees to clients, liabilities, etc. Anyway, even before its termination, the VE may evolve, perhaps to great extent, therefore for the same reasons as for termination, the management of enterprise evolution has also to be explicitly modelled and included into the QMS. Luckily, ISO 9000 standards were meant to be applied to a wide range of enterprises (there are over three hundred thousand ISO 9001 certified organisations worldwide) and thus they are very open. In fact, ISO 9001 clause 5.4.2, item b) states only that the integrity of the quality management system has to be maintained when changes to the quality management system are planned and implemented. The above VE scenario is then fully compatible with the standard.

In ISO 9001:2000, some relevance is given to resource management (clause 6). Differently from to common enterprises the VE does not own human or operation resources, these being modelled as functional entities of the member enterprises. So, at first sight, one could be lead to believe that resource management would be out of scope in the VE environment. However, that is not the case.

Let us first take a look at resource allocation in the VE. Depending on the type of VE operation management perspective, we may encounter different resource allocation strategies:

- *type 1*, where each member makes available functional operations and the associated processing capacity distribution in time, which are used by the VE production planning entity (or entities). Members perform the actual resource allocation to the VE according to its requests.
- *type* 2, where each member makes available functional entities or resources and the associated processing capacity distribution in time, these being used by the VE production planning entity (or entities) to perform the actual resource allocation.

Type 2 must be used whenever the functional entity output is dependent on the entity itself, e.g. artistic or creative work. If type 1 is used, then resource management, from the operations management point of view, is very much simplified by assigning to member enterprises the resources handling tasks. The reverse happens on the quality management system point of view.

When the VE QMS scope was presented (figure 3), it was hinted that members' resources would be included in two (at least) QMSs: the VE's and the member's. Using type 1 resource allocation, all members' resources that may be used in the VE are in fact seen as VE resources. Member's resources responsible for supporting services also have to be included. Unless one is dealing with only a very small part of each member's resources, the obvious conclusion is that the VE's and all members' resource management policies and practices have to be fully compatible.

Type 2 resource allocation may be a little less demanding, from the QMS point of view. In fact, it may be sufficient to monitor only the resources that are "de facto" included and to evaluate if the VE policies and practices are being fully applied. This could well be the case with a VE including a member enterprise with less demanding policies, that allocates a small subset of resources fully compatible to the VE' rules. On the other hand, if the shared resources are a large part of the member' resources then the VE's and the members' resource management policies and practices have to be fully compatible.

Supplier management (item 6.6 in ISO 9004:2000) and purchasing control (item 7.4.1 in ISO 9001:2000) still exist in the VE, being either handled by the VE itself, within the scope of VE operation; or handled by the VE members under their normal operation, but having impact on tasks related to the VE operation. Both types of supplier management and purchasing control related to the VE operation should be fully compatible. The easier way to handle this is to define standard requirements and practices for all members and the VE. In any case, if members do have procurement activities related to the VE operation, then they should be clearly modelled in the VE's QMS, including the supplier evaluation mechanisms.

VE operation and operation control should easily meet quality standards requirements, as the modelling infrastructure that supports VE design and operation requires a complete description of the VE. Nevertheless, extra care must be taken to ensure that every functional entity is qualified to execute its activities and that it has all necessary resources and information. Still regarding VE operation, distinction between a member and a supplier may not be very clear and the smaller the distinction the more agile the VE should be (Goranson, 1999).

From the quality point of view, it is quite possible to soften this distinction, if supplier management is not much different from member management, in what concerns to VE operation. For some VEs (depending on the VE's product/service), it could be possible to define procurement rules and policies applicable to both members and first tier suppliers, as to simplify considerably the VE operation. Such is the case of the automotive industry, where the idea is to push ISO/TS 16949 all the way down the supply chain, ultimately reaching all suppliers. Even so, this is definitely no guarantee of product/service quality (Baiman, 2001).

ISO 9001:2000 clearly states that the QMS should be continually improved (item 8.5.1 *Planning for Continual Improvement*), being this improvement planed and managed according to quality policy, objectives and performance data. The VE's modelling environment proposed above naturally generates in real time important operation data. Customer and member satisfaction data, and other type of relevant information, still have to be gathered by other means (inquires, etc.), as they are not covered by the modelling. Further than quality improvement and the changing client requirements, VE evolution may be determinated by other reasons (e.g. member replacement) and may even result in the decision to terminate the VE operation. Though this is not in even mentioned in the standard, the dissolution of the VE should be planned and controlled.

So far, no big obstacles regarding the implementation of an ISO 9001 compliant QMS in the VE were identified. In fact, the structured VE environment appears to be a natural ground for a QMS implementation. Nevertheless, we believe some requirements regarding VE life cycle should be included in ISO 9001 (or in a complementary standard/guideline).

3. CERTIFICATION

To ensure ISO 9001 compliance of a QMS, a certification body is used (in the USA the certification body is called a *registrar*). Roughly, the process goes through three steps (The European Standard EN 45012 contains general criteria for certification bodies operating quality systems certification). Once a certificate is awarded, there are follow-up audits (usually once a year) and major revision audits (perhaps every three years) to guarantee continued compliance with the standard. ISO 10011 is the international standard used for quality systems auditing (it will be replaced by ISO 19011). The auditing process and requirements defined in this three-part standard is definitely oriented for common enterprise auditing. The VE auditing process must be much more dynamic, as the VE itself is, so that compliance certificate may have any meaning.

We have identified four requirements that may be critical to VE's QMS quality certification:

i. The VE functional entities must meet QMS requirements;
ii. The complete VE operation should be modelled into the QMS;
iii. The complete VE life cycle should be modelled into the QMS;
iv. Continuous auditing of the VE QMS by an independent trustworthy entity (certification body/ registrar) is mandatory.

Items *i* through *iii* result from the above discussion about the ISO 9001:2000 in the VE. An interesting way to meet the first requirement is to have only quality certified members (and suppliers too, if possible), especially if a member is itself a VE. This will not always be possible and procedures for member/supplier management must guarantee the desired product/service quality and preservation along the supply chain.

Item *iv* is perhaps the key to VE certification. It would be a complement to current auditing methods (ISO 10011), but we believe that, due to the agility of VE environment, it is the only way to guarantee continuous compliance with the standard. To achieve this, the VE's operational system encompassing the so-called EMEIS (Enterprise model execution and integration services) will have to allow for the registrar to continuously monitor VE operation, in real time (Figure 4). The VE models should thus provide multiple detail levels, one level being the QMS model (for quality purposes, detailed physical interactions aren't necessary), monitored by the registrar.

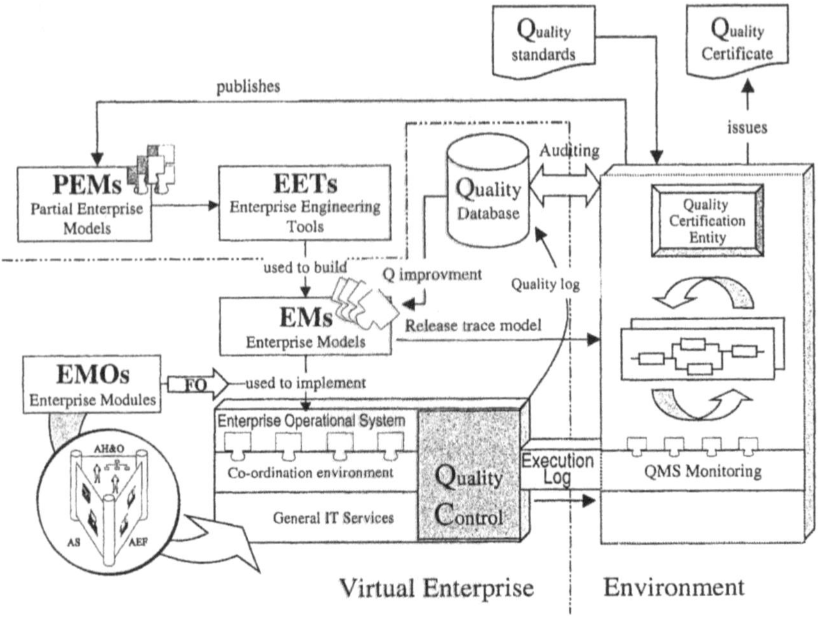

Figure 4 – VE's QMS quality certification scenario

GERAM (IFIP, 1999) introduces the Partial Enterprise Model (reusable, paradigmatic, typical models) concept (figure 4), to capture characteristics common to many enterprises within or across one or more industrial sectors. Thereby, these models may capitalise on previous knowledge by allowing model libraries to be developed and reused in a plug-and-play manner rather than developing every model from scratch. Partial models make the modelling process more efficient. The scope of these models extends to all possible components of the enterprise, such as models of humans roles (skills and competencies of humans in enterprise operation and management), operational processes (functionality and behaviour) and technology components (service or manufacturing oriented), as well as infrastructure components (information technology, energy, services, etc.).

We believe registrars should create partial models modelling the basic QMS infrastructure and its integration across the enterprise with the registrar's. The EV QMS development would thereby be greatly eased (reducing costs and lead time) and these models could even become a differentiating factor between registrars. Representing the particular enterprise, there are the Enterprise Models (EMs), developed by detailing partial enterprise models and introducing other behaviours not modelled in those models. EMs, developed using enterprise modelling languages, include various designs, models prepared for analysis, executable models to support the operation of the enterprise, etc. One of these models is the QMS model, accessible to the registrar for VE operation continuous monitoring and evaluation (figure 4). A Quality Database, storing quality records, models and other EV data deemed necessary by the registrar, is also accessible to the registrar for auditing.

The VE operational system integration with the registrar's is predefined in the Partial Enterprise Model and takes advantage of computer networks, like the Internet. Two levels can be defined (increasing integration):

i. *Quality Auditing*. In this case, a QMS model image may even be stored at the registrar (the quality records must always remain in the enterprise), the registrar having periodic access to the QMS execution log. This log allows for the synchronisation of the QMS model image with the QMS model (through trace driven simulation) and for a complete auditing of the VE (remote access to quality records will be necessary at times). The VE managing entity may be notified of any deviation or nonconformity detected, thus allowing for a better VE management.

ii. *Enterprise Auditing*. A QMS model image may be stored at the registrar or accessed remotely, the registrar having permanent access to the VE QMS, quality records and other company records deemed necessary. The registrar analyses in real-time company data and model execution, and evaluates both quality and performance parameters. This enables real-time reaction to deviation, a feature of critical interest in some cases.

The two monitoring types proposed might even result in different certification levels, e.g. Quality Certification and Enterprise Certification. This mechanism is not only important for VE customers but also for all stakeholders, e.g. VE members, suppliers, shareholders, etc. This concept could even be extended to other areas, such as the Environment and Health & Safety management systems. In fact, environment management system standards (ISO 14000 series) are fully compatible with ISO 9000. There is not yet an ISO standard for health and safety management systems but many national standards are already fully ISO 9000 compatible (e.g. BSI OSHAS 18001).

The Registrar's role is considerable enhanced in this scenario and new revenue sources arise: model construction, continuous auditing and even the chance to move into traditional financial/performance auditing. Finally, for VE quality certification and rating to be used in electronic marketplaces for partner/supplier selection, an adequate security mechanism (using cryptography and electronic signature) must be enforced to guarantee that certificates are authentic.

4. CONCLUSION

In a globalised world, agility and effectiveness are crucial to survive. When technology is allowing for corporations to become agile, the traditional SME's advantages disappear. An interesting emerging solution is the Virtual Enterprise, a low cost operations model to gain size and resources without compromising SME's independence and flexibility.

But for the VE to become a real contender it must break some natural barriers and there is no better answer than to aim for excellence. This is the goal that led us to propose the construction of a VE around a Quality Management System (QMS), including all aspects of VE management and operation. Being the VE a very dynamic enterprise the whole VE life cycle, from creation to termination, should be included in the QMS. This new environment presents some interesting challenges to the quality world. In fact, the evolution of quality from simple inspection to the quality system perspective is being challenged by the extensive use of "variable-geometry" outsourcing. Imposing quality certification on suppliers and then believing that they will smoothly fit into the organisation may be a bit dangerous. Experience shows that putting a group of over qualified people working together does not mean outstanding results will be obtained. We believe a global quality system perspective of the supply chain has to be engineered, and best practices (and requirements if possible) for supplier integration defined. In future work, we expect to address these ideas, specially using some real VEs.

5. REFERENCES

BusinessWeek, "Ford's gamble: will it backfire?", BusinessWeek (European Edition), June 4th, 2001

Camarinha-Matos, L. M., et al, "The virtual enterprise concept, Infrastructures for Virtual Enterprises", Networking Industrial Enterprises, L. M. Camarinha-Matos and Hamideh Afsarmanesh Ed, Kluwer Academic Publishers, 1999

Martins, Angelo, Pinto Ferreira, J. J., Mendonça, J. M., "Quality In B2B E-Commerce: An Objective Tool For Supply Chain Management And Customer Satisfaction", PRO-VE'00, Brazil, 2000

Williams, T.J., "The Purdue Enterprise Reference Architecture" (B-14), H. Yoshikawa and J. Goossenaerts (Editors), Elsevier Science B.V. (North-Holland) 1993 IFIP

Pinto Ferreira, J. J., "From Manufacturing Systems Simulation to Model-based activity co-ordination", CARS&FOF'99, Brazil, 1999

IFIP - IFAC Task Force, GERAM - Generalised Enterprise Reference Architecture and Methodology, Version 1.6.3, March 1999

Goranson, H. T., The Agile Virtual Enterprise - Cases, Metrics, Tools, Quorum Books, 1999

Baiman, Stanley, et al, "Performance Measurements and Design in Supply Chains", Management Science, Vol. 47, January 2001

ISO - International Organisation for Standardisation, http://www.iso.ch

MAKING EFFECTIVE THE INTRODUCTION OF E-BUSINESS IN SME: A REFERENCE MODEL APPROACH

António Lucas Soares[1,2], Luis Maia Carneiro[1], Diana Carneiro[1]
[1]INESC-Porto, Rua José Falcão 110, 4150 Porto Portugal,
[2] Faculty of Engineering of University of Porto, DEEC, Rua Roberto Frias, sn, 4200 Porto, Portugal
asoares@inescporto.pt, lmc@inescporto.pt, diana@inescporto.pt

This paper presents an approach to the identification of requirements for the implementation of e-business solutions in small and medium sized enterprises (SME). The approach is based on the so-called reference models whose main purpose is to guide the process of deriving the detailed requirements for business processes and information systems support, regarding e-business activities. Firstly, an overview of the particular aspects and main difficulties of the adoption of e-business by SME is presented. This overview justifies the methodology for the implementation of e-business in SME, developed within the EC funded project MEDIAT-SME, shortly described here. Next, it is focused a particular aspect of this methodology, the reference models approach, followed by an example of its application on a real SME. Finally, the advantages and difficulties of the approach are discussed.

Keywords: E-Business, SME, E-Business Implementation, Reference Models

1. ADOPTION OF E-BUSINESS BY THE EUROPEAN SME

The past few years provided no definitive answers for the success of the Internet as a new business channel both on the Business-to-Consumer and Business-to-Business fields. Only for very few and specific products it was possible until now to achieve positive results and usually only for big ventures. Even in these specific situations like the selling of books or travel, the long-term profitability is not clear. On the B2B side even though it has been difficult to acquire new customers and implement new business models, there are many known success stories where huge advantages were achieved in terms of reducing costs, reducing time scales and increase quality, with the established business partners. It was also possible to differentiate from competition by providing customers with new services with real added value.

It has been found by Boston Consulting Group (Davis and Gunby, 1999) that the unsuccessful ventures in e-commerce by traditional companies are due to the following factors:

- Failure to do the essential analytical homework,

- Insufficient breadth of vision with respect to potential e-commerce concepts,
- Insufficient breadth of vision with respect to the Internet opportunity beyond e-commerce,
- Delayed attention to fulfilment and customer,
- Inadequate leveraging of core strengths,
- Organisational inflexibility and too little focus on partnering,
- Failure to build in an aggressive "redefine" stage.

The main problem with SME's is that the effort to do the necessary analysis and preparation work is too much. Presently the adoption of e-business by SME's, especially traditional SME's, is very low in percentage terms. The motivation of the MEDIAT-SME project was to allow SME's to enter the digital economy with a reasonable risk and affordable effort from the companies.

2. MEDIAT-SME: A METHODOLOGY

The mission of the MEDIAT-SME project[1] was to design and develop a leading methodology and supporting tools for the efficient introduction of innovative e-Business solutions in SMEs. The main goal of this methodology is to reduce the costs and risks of implementation of e-Business solutions to make it accessible to most traditional SME's. One of the main challenges is to find the optimal balance between the contribution of external consultants and company staff that will allow the minimisation of the total cost while assuring the knowledge acquisition from the company staff necessary to a conscientious decision-making.

The objective was to develop and validate an innovative, application-oriented and practicle methodology and a set of supporting tools covering the analysis of the company's requirements, the specification and design of mediation systems for new types of trade and services, the selection and implementation of adequate ICT solutions in traditional manufacturing & trade SMEs. This methodology strongly takes into account the specific constraints of these types of SMEs in these tasks. Furthermore, it has to minimise the risk for SMEs when introducing new mediation systems and integrating them with the legacy systems. It is expected to bring benefits to the SME in terms of savings in efforts and costs to introduce new systems.

The development of the MEDIAT-SME methodology was based on the analysed requirements with the four end-users of the project consortium, on the requirements of the companies of an interest group and on the large experience of the RTD partners. A business case was defined in each pilot company for requirements definition and validation purposes. The defined Business cases include a service for co-operation between business partners in Engineering tasks for the design of moulds for the shoe industry, a Supply Chain scenario in the building industry and a service of tele-maintenance for producers of manufacturing equipment. A last

[1] IST-1999-11570. http://www.cbt.es/mediat-sme

Generalisation Business case was defined to assess the applicability of the developed methodology to other industries. The basic concept of the MEDIAT-SME Methodology is based upon four key principles:

1. "Business objectives satisfaction": The methodology assures that the selected ICT supported solutions contribute to the realisation of the organisation's business objectives.
2. "Bottleneck identification": The analysis identifies the critical bottlenecks of the SMEs, by comparing the company practices with a defined reference model, which provides the Best Practices for different industrial sectors.
3. "New opportunities identification": The main procedures and structure of the Methodology is primarily oriented to introduce improvements in the SME and its relations with business partners. In this scope potential e-Services corresponding to the identified bottlenecks have to be analysed for each customer group. A further essential point is the identification of new market opportunities posed by the internet.
4. "Phased approach": The overall life-cycle to introduce new ICT supported solutions has to follow a sequence of phases starting with the analysis and selection up to the implementation. Each phase is performed on a specific level of abstraction. The methodology strongly supports a phased process for the introduction of ICT systems and corresponding organizational changes.

The Methodology concept (see Figure 1) divides the life cycle of introduction of innovative e-Business solutions in three main phases: Analysis, Design and Implementation. Two support components have also been developed providing training materials on e-Business and on the methodology itself and an innovative Building Block Library of software components covering functionality's not addressed by market available solutions. The main objectives of the Analysis phase are the identification of bottlenecks, the diagnose of the relation practices of the company with its clients and suppliers, identification and analysis of improvement opportunities and definition of an implementation concept.

By the end of the analysis phase the management of the company should have acquired the necessary information to decide the implementation of a particular solution concept. This decision is based on a Return On the Investment analysis.

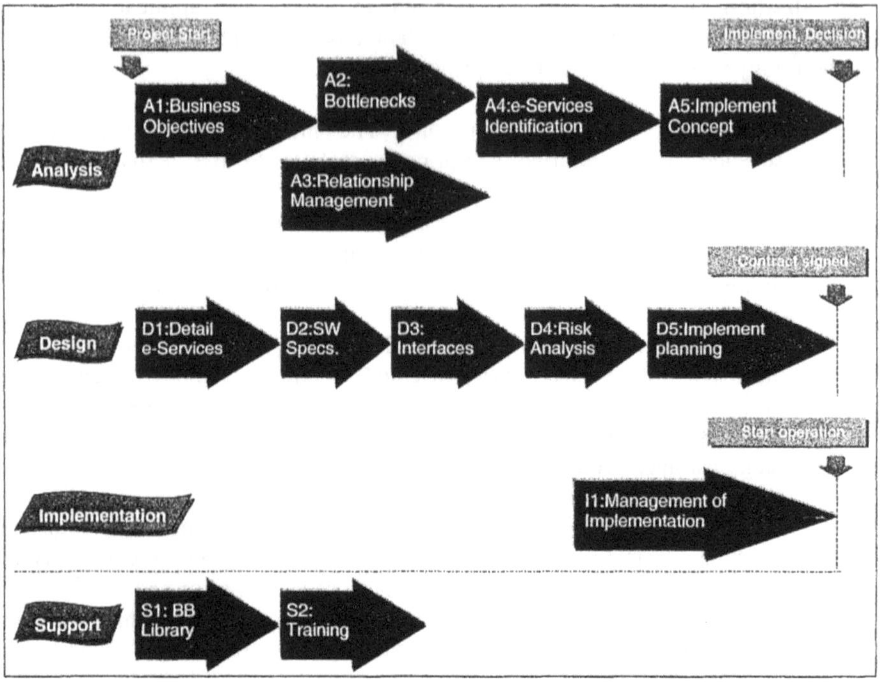

Figure 1 - Overview of the MEDIAT-SME methodology

One possible result of this phase is the convincement of the project team that there are no solutions worth to be implemented. In this scenario, the following phases of the methodology won't be implemented.

In the second phase of the Methodology the support organization, software applications and interfaces with legacy systems are designed. The software specifications are done just at the detail level necessary to allow software vendors to perfectly understand the company requirements and give a quotation. A risk analysis is conducted. Finally, a detailed implementation plan is developed. A Request For Proposals is elaborated, which includes the software specifications, system architecture, SLA, etc. This phase includes the selection of the best offer and is finished with the signature of the contract with the software vendor(s) selected.

The implementation phase supports the management of all implementation process, including the development, integration, verification, quality, logistics, marketing, organisational change and training aspects. The implementation shall follow the detailed implementation plan defined in the design phase. In this phase the MEDIAT-SME provides detailed guidance on how to manage the implementation of e-Business projects.

3. REFERENCE MODELS

3.1 Business process modelling

Business process (BP) modelling is a relatively new area even thought that some modelling concepts where established more than two decades ago. In the last few years, the business process focus of enterprise engineering together with the expansion of the e-business paradigm demanded for effective enterprise modelling methods. Before proceeding with the explanation of the reference models approach, we will expose our view of the concept of business process and BP modelling[2]. Business process is an inter-related set of activities, partially ordered by precedence relationships, triggered by some event(s) and producing some observable result (Vernadat, 1997). A business process model is a conceptual representation of a BP as viewed from several perspectives, the ones needed to represent the BP adequately for a given purpose. Business process modelling is the process of building and maintaining BP models.

The main purposes of business process modelling are twofold, although overlapping: (1) to support the management of the business processes and (2) to support their design and implementation. The first purpose involves the documentation facet of modelling, providing the people involved in managing and executing the BP activities with a tool to (hopefully) understand better their work. The second purpose has to do with the (re)design of BP, supporting people in the enterprise in setting up new or redefined BP more efficiently. This includes obviously the documentation facet. Regardless the main purpose of the BP models it is necessary to decide on the perspectives that are necessary to consider in order to represent adequately a BP. This is usually called modelling "views". Therefore, the adoption of a given modelling framework depends very much on the BP aspects that must be analysed and managed with the support of models. There are a number of modelling frameworks or (modelling methods) that provide a comprehensive set of tools from several modelling views. It is not a goal of this paper to detail this issue (regarding enterprise modelling methods in general see Kingston et al., 1994; regarding the organizational view of enterprise modelling methods see Soares, 2000; for a specific framework see for example Scheer, 1999; on business process modelling see Becker *et al.*, 2000).

Within the broader context of our R&D activities the UML modelling method was adopted as standard. Regarding BP modelling a complementary method involving specific extensions was adopted (Eriksson and Penker, 2000). UML includes a set of well known and easy to learn diagrams that, together with established and/or customised procedures, enable a representation of the business processes adequate to several types of analysis. Another important issue is the object-oriented feature of UML. Among the characteristics of the object-oriented perspective, one should highlight here the generalisation/specialisation flexibility,

[2]This is needed because these concepts are interpreted with some inconsistencies by the research and industrial communities, leading to misunderstandings when defending a particular approach. Nevertheless, the definitions presented here are not intended to be the ultimate definitions, but their purpose is only to make clear the arguments in this paper.

enabling the construction of models with a variable level of detail. This can result in models generic enough to be the reference for several companies from one side, and to be easily specialised for each application from the other side. This perspective pervades all the UML diagrams and is well suited to the purpose of reference models.

3.2 The basics of the e-Services Reference Models approach

Reference models in the MEDIAT-SME methodology are a guide for e-services specification, implementation and maintenance. Reference model here means a model that represent a more or less consensual description of a given BP. It consists in a set of UML diagrams reflecting standards, best-practices, experience and new approaches regarding BP including e-business components. The description conveyed by the reference model should be at a level of generality that enables its application to biggest number of SME's. These models address mainly the specification phase and are the input for the implementation phase of the methodology.

The first version of e-SRM includes four UML diagrams: process diagrams (Eriksson-Penker extension, Eriksson and Penker, 2000), activity diagrams, class diagrams and use-case diagrams. The first three are needed for a minimal description of the BP's involved in the e-Service and the forth is a first and rough specification of the information systems support to the BP's. A process diagram is an high level description of a BP. It is inspired in IDEF0 diagrams depicting inputs, outputs, constraints and resources. It adds the concept of BP objective, truly important to analyse the "why?" of the BP. Process diagrams can be decomposed, introducing the concept of sub-process and enabling the hierarchical organization of the process diagram. Activity diagrams describe the sequences of activities in a (sub) process with no further decomposition. An activity diagram can be enriched with object flows, showing the dependencies between data objects and activities. Class diagrams describe the basic concepts in the BP's domain, their relationships and organization, being the first iteration for a data model.

The mode of use of the e-Services Reference Models (e-SRM) should follow the general object-oriented perspective in UML. The e-SRM should be used less in a prescriptive way and more in a way of questioning and reflecting about the needs of each company, regarding the specific e-services. It is our opinion that it is very difficult to come out with a recipe for the implementation of such services. This would result in problems in the analysis of the situation (consultancy phase) or (even worst) in problems in the implementation phase. Therefore, we envisage the use of the e-SRM as a kind of sophisticated "check-list" that enable the company and the consultant to go step-by-step in the construction of a detailed e-service specification that should be the most adequate for the situation in hand.

4. AN EXAMPLE

The application of the Analysis phase of the MEDIAT-SME Methodology to the identified pilot company suggested that the e-services which could bring relevant benefits to the company where related to after sales support services (*e-after-sales*) and *e-learning*. The chosen pilot company manufactures and sells highly complex and critical production equipment, which have demanding requirements in terms of maintenance and technical assistance and support. It was defined an implementation concept including tele-maintenance, technical assistance and e-learning. The expected benefits in terms of sales of maintenance and support services and reduction of operational costs in this area will allow the Return On the Investment to be archived 2.1 years after the start of the operation. The example described in this section focus on the specific business process named "Maintenance Management" and its sub-process "Remote Maintenance Management" (RMM).

4.1 RMM as part of e-After-Sales service

Remote maintenance management (RMM) is the most effective solution to provide a high quality of service at an adequate cost, and can constitute an effective strategy for the Portuguese equipment supplier companies to overcome competitive weaknesses. Nowadays, there are few Portuguese equipment manufacturing and seller companies that provide remote technical assistance and maintenance to their customers. Furthermore, few SME's see e-Bussiness concepts as important for an after-sales strategy. A RMM system should support a set of maintenance and technical assistance services to be carried out remotely by the equipment supplier companies. For that, this system must provide a set of functionalities to detect, diagnose and correct malfunctions or breakdowns remotely. This implies, of course, the remote access to the equipment controllers and data. As basic requirements, a RMM system should: monitor remotely of the equipment status, detect and diagnose malfunctions and breakdowns, store breakdown, malfunction and repair data, notify maintenance managers, visualise the status of the monitoring components, visualise maintenance state variables, control the equipment remotely, remotely access to the equipment technical data, extract and visualise performance indicators regarding interventions and maintenance costs, assist technically in the of equipment problems. Additionally, RMM services must enable the co-ordination with local maintenance activities. RMM cannot cover all the maintenance needs, but it should support complementary actions performed on-site by the company technicians.

4.2 Application of the e-After-Sales RM to the RMM case

Figure 2 shows an extract of the e-After-Sales RM. Figure 2a is the top-level business process model. Figure 2b is the detail of one of the sub-processes ("Maintenance Services") that compose the e-After-Sales BP. This process model also shows the decomposition of the "Maintenance Services" BP into the lower level sub-processes. One of these lower-level sub-processes is represented in figure 2c, being its activity model described in the diagram of figure 2e. Finally, an extract of the use-case diagram of an information system supporting the sub-process of "Maintenance Request and Planning" appears in figure 2d.

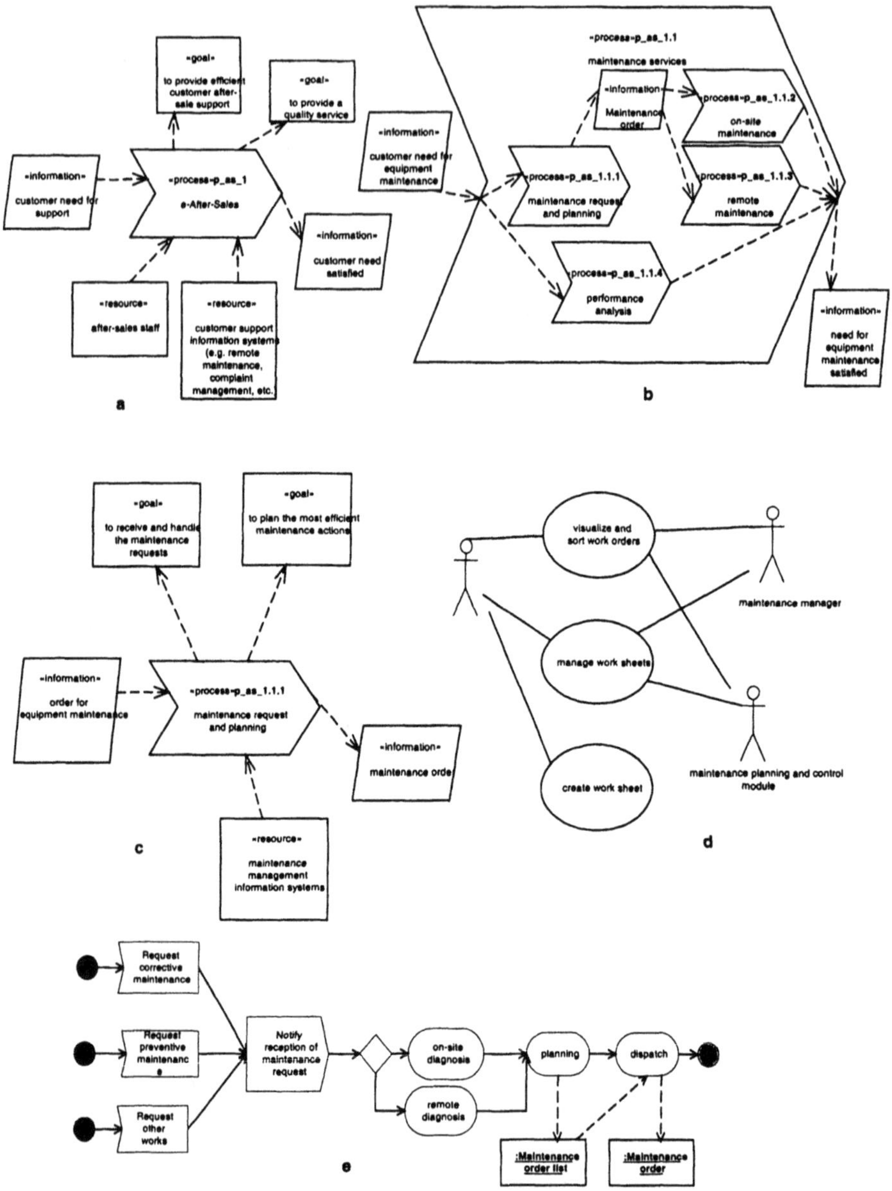

Figure 2 - Extract of the e-After-Sales RM

The company where this analysis was done was already in the way to implement a concept of remote maintenance. The e-SRM method enabled a structured approach to this task, enforcing the particular set of RMM activities to be analysed and set-up in the broader context of the After-Sales business process. Regarding the specialization of the e-SRM, in this particular example models 2c, 2d, and 2e where slightly changed in order to accommodate company's specific requirements.

5. CONCLUSION

The concept of "Reference Model" is used in theory and practice with different meanings and origins distinct practical approaches. A RM has a prescritive nature that can cause difficulties when applied to practical cases. Besides problems such as the customization of the models or the complete unadequacy to some situations, it can be the case of not motivating enough reflexion on the analysis of the BP in the company. The approach suggested in this paper tries to overcome some rigidity associated with the concept of RM. One the one hand, e-SRM adopt a generality level that is neither to abstract (leading to vagueness), neither to detailed (leading to rigidity or making necessary too much changes). On the other hand, the adoption of the UML business modelling method with the underlying object-orientation makes conceptually easier to derive specific models adequate to the situation in hands.

This approach was tested in one company with promissing results. Further work will be to apply it to more practical cases collecting practical experience crucial to refine the whole approach. This is being made in consultancy projects using the MEDIAT-SME methodology.

6. REFERENCES

Becker J, Rosemann M, Uthmann C., 2000, Guidelines of Business Process Modeling. Business Process Management: Models, Techniques, and Empirical Studies, van der Aalst et al., ed. Berlin: Springer.

Scheer, A-W., 1999, Business Process Modeling. Berlin: Springer.

Davis, J., Gunby, S., 1999, Winning on the Net: Can Bricks-and-Mortar Retailers Succeed on the Internet? Boston Consulting Group.

Eriksson, Penker, 2000, Business Modelling with UML. Addison-Wesley.

Soares, A. L., 2000, A social and organisational ontological foundation for Enterprise Modelling, in IFAC Symposium on Manufacturing Modelling, Management and Control (MIM 2000). Patras, Greece.###

Kingston et al., 1994, Enterprise State of the Art Survey - Part 3 - Enterprise Modelling Methods, Ent/DE/3/1.0, University of Edinburgh.

Vernardat, F., 1996, Enterprise Modelling Methods - Principles and Applications, Chapman&Hall, London.

PART THREE
MECHATRONICS

ROBOT CALIBRATION USING GENETIC PROGRAMMING

Jens-Uwe Dolinsky
Gary Colquhoun
Ian Jenkinson
Liverpool John Moores University, UK.
Corresponding author: G.J.Colquhoun@livjm.ac.uk

Conventional robot calibration techniques rely on numerical optimisation methods with the attendant problems of; kinematic equation non-linearities, inappropriate model parameterisations and parameter discontinuities or redundancies. In this paper research using a symbolic co-evolutionary calibration algorithm is described. The approach merges established kinematic modelling techniques with Genetic Programming to generate joint correction functions as part of an inverse calibration model. The data generated in a calibration procedure is described and the paper concludes by discussing the potential for symbolic calibration and the automatic generation of correction models using Genetic Programming.

Robot calibration, Genetic programming, Kinematic model, positional error, symbolic regression

1 INTRODUCTION

Industrial robots are widely used to automate tasks such as assembly, welding, paint spraying and machine loading/unloading. Improved control systems, better off-line programming capability, and reductions in initial purchase costs are opening more opportunities for their use in a global commercial environment geared to cost reduction and quality conformance.

Robot task programming is largely carried out using manual teach programming with the robot out of production. As task complexity increases this becomes time and therefore cost intensive and Offline-Programming (OLP) systems promise to remove many of the drawbacks inherent in conventional teach programming. An OLP system uses a robot model to create a virtual environment in which robot tasks can be simulated with the potential to: use imported CAD data to simplify path geometry, to reduce plant downtime and programming time, to de-bug, optimise and validate paths and to eliminate safety problems. Examples of OLP systems include: Workspace®/DELMIA/IGRIP® (DELMIA, 2000), FAMOS®(1998), RobotStudio® (ABB, 2000) and Ropsim® (CAMELOT, 2000).

Despite the potential for OLP systems they are limited by inherent robot arm positional inaccuracy and by the inexact modelling of those inaccuracies in robot controllers. The difference between a simulation model in a virtual environment and the real robot can cause OLP systems to generate robot poses with positional errors.

Additionally the path-planning algorithms used in OLP systems are often different or simplified versions of those used by the robot manufacturer which can result in unreliable cycle time predictions, collisions and sub-optimal end effector paths. This research is concerned with static calibration and the identification of accurate models of physical properties and effects that influence the static positional accuracy. Static calibration involves: the development of a suitable kinematic model providing a model structure and nominal parameter values, measuring end-effector location in several positions, the identification of the model parameters using numerical fitting based on actual end-effector position measurements.

2 ROBOT CALIBRATION

Assuming that a robot is working at its specified mechanical accuracy any accuracy improvements rely on software based approaches. The *task space compensation* approach subtracts the expected positional error (derived from an accurately calibrated model) from the target pose. In this way false target poses are produced and the deviation of the manipulator at these altered poses leads the robot to the desired position. To deal with inaccuracies offline generated poses need to be corrected using an accurate kinematic model. Conventional calibration techniques rely on numerical optimisation methods for model parameter estimation. However, kinematic equation non-linearities and inappropriate model parameterisations, with possible parameter discontinuities or redundancies, can lead to badly conditioned parameter identification. Research in kinematic calibration has largely focused on finding robot models and appropriate numerical methods to increase accuracy. Most work in this area has reported calibration applied to the forward kinematic model $\mathbf{y} = \mathbf{f}(\theta,\phi)$ such as (Ambarish et al, 1993), (Hollerbach and Wampler, 1996) and (Judd and Knasinski, 1990). The inverse kinematic calibration procedure (figure 1) attempts to identify the parameter vector ϕ using the inverted kinematic model $\theta = \mathbf{f}^{-1}(\mathbf{y},\phi)$ Inverse calibration is in general more difficult because it requires the model $\mathbf{f}$ to be analytically invertible, which may not be the case with very complex models and models of redundant[1] manipulators.

Figure 1 - Task space compensation in OLP using a nominal inverse kinematic model

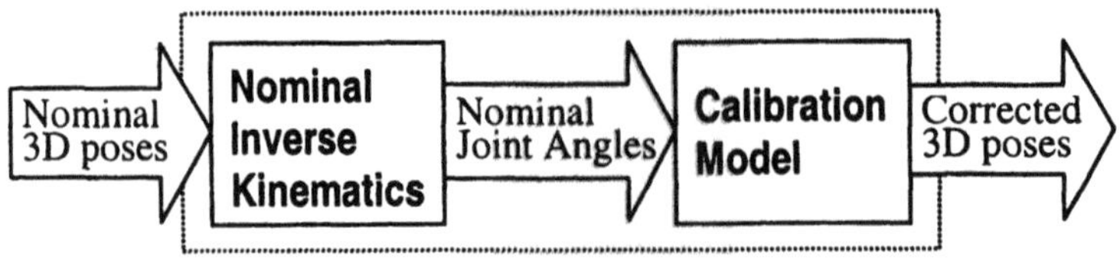

A kinematic model can be defined as *proportional* if small changes of the physical properties can be represented by small changes of related model parameters. A kinematic model is said to be *complete* if all kinematic properties of

[1] Redundant manipulators can reach a given pose using different joint configurations.

the manipulator are represented by corresponding independent model parameters. In this way all possible kinematic configurations of the manipulator can be sufficiently described if small changes can be represented by small changes of related model parameters and all kinematic properties of the manipulator are represented by corresponding independent model parameters. For an incomplete kinematic model the number of model parameters is usually smaller than the number of kinematic properties of the manipulator. In this case parameters account for modelled properties as well as non-modelled effects. An identification algorithm might be able to find optimal parameter values, however these values will be optimised for this particular incomplete model and may not reflect the physical properties of the robot. In order for a model to be complete it should describe both geometric properties (link length and twist, joint encoder offsets) and non-geometric effects such as gear backlash, static deflection and encoder errors.

Errors in geometric parameters are claimed to contribute most positional inaccuracy (Caenen and Angue, 1990) (Judd and Knasinski, 1990). Although less significant, the positional error of non-geometric effects cannot be neglected if high positional accuracy is required. However they are more difficult to model as they vary with manipulator pose and payload (Vincze et al, 1996).

3 KINEMATIC MODELLING

The most popular method of modelling robot kinematics is serially composing link models using the standard Denavit-Hartenberg (DH) parameterisation. DH modelling is relatively easy to use and often yields an algebraic inverse solution, which can be quickly computed. However, it has been shown to be unsuitable as a model for robot calibration. The kinematic models do not have enough parameters to express any small variation in the actual robot structure away from the nominal. Small variations in the actual robot structure do not result in correspondingly small variations in model parameters. For example large changes in the model parameters occur when consecutive joints change from parallel to almost parallel planes. The result is instability in the numerical methods used during the identification phase. Also world and tool frames cannot be located arbitrarily and variations in robot geometry will therefore result in variations in the position of these frames. The constraints of DH parameterisation have been tackled by many authors (Ambarish et al 1993), (Hayati and Mirmirani, 1985), (Hollerbach and Wampler, 1996), (Stone,1987), (Zhuang et al, 1992), (Mooring et al, 1991), (Everett,1993), (Vincze (a), et al, 1994).

Non-geometric effects are usually modelled by adding terms or components to the overall geometric model of the manipulator. Since non-geometric effects are primarily due to joint related characteristics (Vincze (b) et al, 1994) the most common model adopted is a simple linear joint correction model: $\mathbf{\theta} = \mathbf{k}\theta + \gamma$ applied to each joint. Research in this area has developed methods such as the use of different models for each joint (Whitney et al, 1986) and the use of Fourier series to approximate non-geometric effects (Everett, 1993), (Vincze, 1996).

The complexity of positional error leads to error approximation rather than explicit definition using parametric models. Functional approximation provides a variety of approximation models and methods based on uni- and multivariate polynomials, splines, Bezier curves, wavelets, Fourier series and artificial neural networks (Lewis et al, 1996) (Jenkinson, 2000) where an approximation model is fitted (or trained) to the data. However, approximation models are inherently non-parametric, i.e. there is no semantic relationship between model parameters (or coefficients) and physical properties. These models are valid only in the interval they have been trained for and are weak or not capable of extrapolating or generalising beyond these interval boundaries.

4 SYMBOLIC CALIBRATION

The application of symbolic model regression to the problem of static kinematic calibration is a significant development over numerical based methods. The approach proposed removes the problems of ill-conditioning and local convergence by merging established kinematic modelling techniques with Genetic Programming (GP). In this work it is used to generate joint correction functions as part of an inverse calibration model. Unlike conventional calibration methods the process is automatic, it does not start with a predefined model. Instead the calibration model is built up from primitive model components during the calibration process. Since there is no iterative numerical parameter identification involved, corresponding stability and conditioning issues are of no concern.

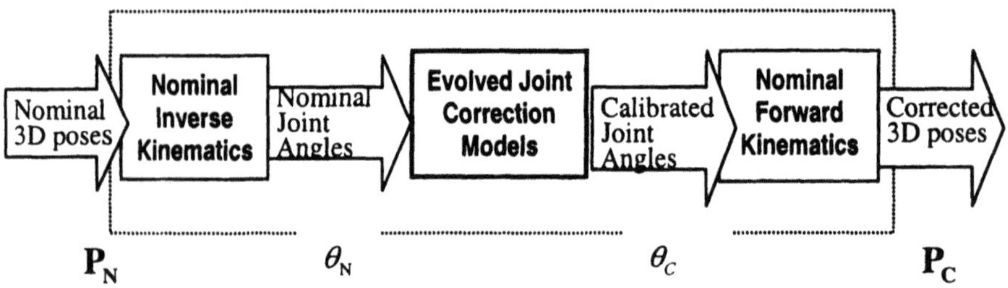

Figure 2 - The co-evolutionary calibration method

Figure 2 illustrates the overall calibration method where: $\mathbf{P}_N$ is the nominal end-effector pose, θ the calibrated joint configurations and $\mathbf{P}_C$ corrected end-effector poses. The task is to create an accurate mapping between nominal and corrected end-effector poses by establishing the required calibrated joint configurations. Generating calibrated joint configurations from nominal joint configurations is accomplished by evolving joint correction models implicitly using a forward kinematic model as a distal teacher (Jordan and Rumelhart 1992). The evolution of the correction models is driven by their distal performance i.e. the accuracy of the whole forward kinematic model (geometric model + evolved joint correction models). The distal supervised learning approach is used to establish joint correction functions, which model the error of the nominal inverse kinematic model. The evolution of these functions is performed sequentially using a co-evolutionary method (shown in general terms in figure 2) to minimise the performance index: a measure of the total positional error of the robot end-effector.

After creation and initialisation of the population, the components of the correction function vector **g** are initialised with neutral functions. This ensures that the initial performance index (the measure of error at the beginning of the evolution) of the kinematic model is equal to the performance index of the uncalibrated kinematic model (without correction functions).

An evolutionary cycle (the *repeat-until* statements in figure 3) starts by selecting a joint corresponding to the contribution of its joint correction model to the positional error. The forward kinematic model is represented by a sequence of homogenous DH transformations. Hence the joint correction models (which are arguments of their corresponding DH transformations) collectively contribute to the performance index within this serial arrangement. This also implies that the correction models depend on each other during the evolutionary learning process. When altering one correction model the other models are directly affected in their potential to minimise the reminder of the positional error. Since this property prohibits a parallel implementation of evolution for each joint only one joint correction model is evolved at a time. The objective to minimise the performance index in the "environment" provided by the other correction models. This is co-evolution since the performance of individuals depends on the performance of individuals in other populations. While one correction model is evolved in its corresponding population, the evolution of the other models is suspended. The generations used by the calibration system to evolve the correction function vector are the sum of the generations produced by each step in the genetic programming algorithm.

k:= Number_of_joints
P:=Create populations of correction functions for **k** joints
for each k do
    $\mathbf{g}_k(\theta) := \theta$
Compute initial performance index
repeat
    j := Select Joint according to the contribution of its corresponding correction model in g to the positional error of the end-effector
    Evolve correction model for joint **j** in population **P[j]** until performance index improves
    Update $\mathbf{g}_j(\theta) :=$ $\mathbf{g}_j(\theta) :=$ fittest model of population **P[j]**
until terminated

Figure 3 - The symbolic co-evolutionary calibration algorithm

Whenever the performance index (and thus the positional error) has been improved (which may take several generations) by evolving a particular joint correction model, the next step in the algorithm is to update the corresponding component in the correction function vector **g** with the fittest model of that population. Then the evolutionary cycle starts again by selecting a joint based on the contribution of its correction model in the now updated function vector **g** to the remainder of the positional error. This evolutionary cycle is repeated until the performance index drops below a certain limit or the sum of generations used to evolve all joint correction models exceeds a certain number.

5 DISCUSSION

The results from the initial steps of a typical run (figure 4) illustrate the calibration procedure. Rows contain the evolved correction model of a joint, the generation at which this correction model emerged and the resulting improved performance index.

Joint number	Gene-ration	Best correction model	Perform. Index
2	3	theta+COS(SIN(theta))*0.3842585528122807%SIN(0.0805383465071 5659+theta)*SIN(SIN(0.008484145634327219))*0.08981597338785974	32.9893
2	5	theta+COS(SIN(theta))*0.3842585528122807%SIN(0.05172887356181524+theta)*SIN(SIN(0.008484145634327219))*0.08981597338785974	31.5853
2	8	theta+COS(theta-0.6748252815332499)*0.3842585528122807%SIN(0.08053834650715659+theta)*SIN(SIN(0.008484145634327219))*0.08981597338785974	30.3445
1	10	theta*COS(0.02633747367778558*(0.3003936887722404-theta)*0.07809686574907682)	30.3441
1	11	theta*COS(0.02002014221625416%SIN(theta))	29.9366
1	14	theta*COS(SIN(0.9015778069399091-0.4301278725547044*theta*theta)*0.07809686574907682)	29.1421

Figure 4 - Symbolic expressions of the correction models generated (initial performance index of 37.9771)

Joint 2 is first selected. Breeding of new correction models in the corresponding (second) step in the GP algorithm is performed until the performance index could be improved in generation three. The second component in the correction function vector is updated with the remainder of the positional error and the updated correction vector joint 2 is selected again for further correction model improvement. In this way the GP algorithm gradually improves the performance index by refining the correction model of the second joint. In generation eight however a correction model emerges which reduces the positional error in a way that gives a correction of joint 1 more potential for a further reduction of the performance index. Hence from generation 9 the algorithm switches to joint 1 and evolves the first correction for this joint in generation 10 with further refinements in generation 11 and 14. In this way the symbolic calibration algorithm automatically selects and improves joint correction models.

These initial results have demonstrated the potential for symbolic calibration and for the automatic generation of the correction models. Some limitations have been encountered in terms of computing resources and time and convergence speed in later generations and in the unpredictability of the progress of evolution. Research is being pursued to using parallel processing and the application of advanced evolutionary research.

6 REFERENCES

ABB. RobotStudio – Programming & Simulation. RobotStudio® product description. Hannover. 2000.

DELMI. IGRIP Reference Manual. DELMIA Corporation. 2000.

Ambarish G, Quaid A, Peshkin. M. Complete Parameter identification from partial pose measurement. IEEE International Conference on Robotics and Automation. 1993.

Caenen J, Angue J. Identification of geometric and non-geometric parameters of robots. Proceedings of the IEEE International conference on robotics and automation. 1990; 1032-1037.

CAMELOT. Ropsim product description. Allerød. Denmark. 2000.

Everett LJ. Models for Diagnosing Robot Error Sources. Conference on Robotics and Automation. 1993; 2: 155-159.

FAMOS® *FAMOS* Product description. Carat robotic innovation GmbH. 1998.

Hayati SA. Mirmirani M. Improving the absolute positioning accuracy of robot manipulators. Journal of Robotic Systems. 1985; 2: 397-413.

Hollerbach JM. Wampler CW. A taxonomy of kinematic calibration methods. International Journal of Robotics Research. 1996;14,: 573-591.

Jenkinson, I.D. *An application of neural networks to improve the accuracy of an industrial robot for offline programming*. Ph.D. Thesis, Liverpool John Moores University. Liverpool.. 2000.

Jordan MI. Rumelhart DE. Forward models: Supervised learning with a distal teacher. Cognitive Science, 1992; 16: 307-354.

Judd RP, Knasinski AB. A Technique to Calibrate Industrial Robots with Experimental Verification. IEEE Transactions on Robotics and Automation. 1990; 6(1): 20-30.

Lewis J, Zhong X, Nagy FN. Inverse robot calibration using artificial neural networks. Engineering Applications of Artificial. Intelligence. 1996, 9: 83-93.

Mooring BW, Roth ZS, Driels MR. Fundamentals of Manipulator Calibration. John Wiley & Sons. 1991.

Stone HW. Kinematic Modelling, Identification and Control of Robotic manipulators. Kluwer Academic Publisher New York. 1987.

Vincze M (a), Prenninger JP, Gander H. A laser tracking system to measure position orientation of robot end effectors under motion. International Journal of Robotics Research. 1994; 13: 305-314.

Vincze M (b), Filz KM, Gander H, Prenninger JP, Zeichen GA. Systematic Approach to Model Arbitrary Non-Geometric Kinematic Errors. Advances in Robot Kinematics and Computationed Geometry. Kluewer Academic. 1994: 129-138.

Vincze M, Filz KM, Gander H, Prenninger JP. A Systematic and Extendible Model for Arbitrary Axis Configurations. International Journal of Flexible Automation and Integrated Manufacturing. 1996; 3(2): 147-164.

Whitney DE, Lozinski CA, Rourke JM. Industrial Robot Forward Calibration Method and Results. Journal of Dynamic Systems, Measurement, and Control. 1986; 108: 1-8.

Zhuang H, Roth RS, Hamano F. A complete and parametrically kinematic model for robot manipulators. IEEE Transactions on Robotics and Automation. 1992; 8,(4).

ROBUST CONTROL OF AN ADAPTIVE OPTICS SYSTEM USING H_∞ METHOD

Benjamin West Frazier*
Robert K. Tyson*
Y. P. Kakad**
B. G. Sherlock**
** Department of Physics*
*** Department of Electrical and Computer Engineering*
University of North Carolina at Charlotte
Charlotte, NC28223, USA
bwfrazie@earthlink.net, kakad@uncc.edu, sherlock@uncc.edu

Adaptive optics systems are utilized in systems, which require the real-time correction of blurred images and have applications in astronomical imaging, free-space optical communications and high-energy laser beam guidance. This paper discusses the proposed design for an adaptive optics system that utilizes modern H_∞ control theory to optimize the design of this multivariable control system in terms of the stability and performance robustness of the overall system.

Keywords: Adaptive optics, multivariable controls, optimization, image enhancement, filtering.

1. INTRODUCTION

Adaptive optics systems have received a great deal of attention because of their ability to correct aberrated beams of light. In the most recent news, a brown dwarf star previously invisible was imaged using adaptive optics at the Gemini North and Keck observatories in Hawaii (Michaud, 2002). The use of adaptive optics in this case allowed researchers to produce very sharp images of the companion star which is too faint and too close to its parent star to be seen otherwise.

This attention provides critical acclaim for adaptive optics. However, both military and research applications demand continually improving performance from these systems. Therefore, a practical adaptive optics system should be optimized to attain its best possible performance. There is a wealth of research available in the literature for optimizing adaptive optics systems, but these are generally focused on optimizing only the wavefront estimates (Hudgin, 1977; Irwan, 1999; Paschall, 1993; Wallner, 1983), or the individual control loop bandwidths (Ellerbroek, 1994). The most promising methods today for real-time control involve adaptive techniques such as those outlined in (Ellerbroek, 1998).

The fundamental problem with most of these optimization techniques is that they assume almost perfect knowledge of the process to be controlled. In an adaptive optics system, this would mean that we have complete *a priori* knowledge of the atmospheric turbulence where the system is operating. This assumption is not always valid, as in many practical cases we have incomplete knowledge of the system, which translates to uncertainty in our model. For example, adaptive optics systems are sensitive to variations in altitude, temperature and atmospheric turbulence and we cannot always predict the conditions under which the system will operate. Another problem that arises with most modern techniques is that the non-stationary nature of atmospheric turbulence forces us to recalculate the control matrix every few minutes and to use control scheme such as gain scheduling. In a similar fashion, a system can be moved or modified for a different set of conditions and then must be re-optimized. This paper presents a method to deal with these uncertainties by utilizing H_∞ techniques which optimize the robustness of the system under those uncertainties rather than optimizing the system performance in a given environment.

To begin this paper, we first provide an introduction on the basics of an adaptive optics system, and then discuss what we mean by H_∞ filtering and control. Finally, we demonstrate how the wavefront estimation process can be improved by using H_∞ filtering techniques.

2. ADAPTIVE OPTICS SYSTEMS

Introductions to the field of adaptive optics and detailed descriptions of working systems may be found in (Hardy, 1998) and (Tyson, 2000). In this paper, we will only discuss the most basic concepts at the system level.

An adaptive optics system generally consists of three main components: the wavefront sensor, the control computer and the deformable mirror. The wavefront sensor determines the phase of an optical wavefront and is typically made up of a CCD camera and specialized software that calculates the wavefront estimates. The control computer calculates the corrections to be made and is the heart of the system. Finally, the deformable mirror implements the corrections. A deformable mirror is usually a thin facesheet with either electrostrictive, (e.g. PZT), or magnetostrictive, (e.g. PMN), actuators attached (Tyson, 2000). The deformable mirror is often characterized by its influence function matrix, which is the relationship of the actuator commands to the mirror surface (Tyson, 2000). In a feedback control loop, the wavefront sensor can be thought of as a state estimator, while the plant we are trying to control is the optical wavefront itself. The deformable mirror then is the control element and the control computer ties everything together and implements the algorithm.

The state-space model for a typical discrete-time system is given as the following (Franklin, 1990):

$$x_{k+1} = Ax_k + Bu_k + w_k \tag{1}$$

$$y_k = Cx_k + v_k \tag{2}$$

where x_k represents the current state, u_k is the current input command, w_k is the process noise (disturbance), y_k is the current output, v_k is the measurement noise, and A, B and C are the system matrices.

From (Furber, 1997), we can write the state-space realization for an adaptive optics system as the following:

$$\phi_{\rho,\theta,k+1} = \Gamma \phi_{\rho,\theta,k} + Z_{\rho,\theta}[a_k + d_k] + v_k \tag{3}$$

$$y_k = S\phi_{\rho,\theta,k} + n_k \tag{4}$$

where y is the output vector, d is the input disturbance vector at the deformable mirror, v is the input disturbance vector at the wavefront, a is the vector ofactuator commands, Z is the matrix of influence functions, n is the vector of measurement noise added at the wavefront sensor, S is the wavefront sensor operator, Γ is a constant matrix that weighs the importance of the previous states, ρ and θ are the circular coordinates defined over the optical aperture, and ϕ is the state vector of wavefront aberrations at discrete points over ρ and θ.

It is clear that equations (3) and (4) are in the same form as equations (1) and (2), so utilizing this model, we can proceed with standard control and estimator design techniques. The critical point is to determine the system matrices Γ, S, and Z which may be derived by finite element analysis or interferometric measurements (Tyson, 2000).

3. H_∞ FILTERING AND CONTROL

H_∞ control theory is a highly mathematical approach to modern control that seeks to optimize system performance in the presence of uncertainty. In the theory, we utilize a special set of function spaces, known as Hardy spaces with the property that all of the functions in a Hardy space are proper, rational and stable transfer functions. With the wealth of knowledge in the fields of linear operator theory and algebraic topology, H_∞ control theory was developed so that we can force the system transfer function to be a member of a Hardy space, regardless of the uncertainties present (Zhou, 1996). This condition allows us to ensure that our system remains stable under any operating conditions.

The H_∞ filtering problem can be stated as an attempt to find the state estimate that minimizes the worst possible effect that the input noises have on the estimation error. As a result, the H_∞ filter achieves the most robust performance possible. This differs from the typical optimal estimation problem which seeks to minimize the variance of the estimation error to achieve the best possible performance. If we let J be a cost function, or a measure of how good the estimate is, we can attempt to minimize this cost in the presence of worst case noise. Unfortunately, we cannot find a closed form solution for this, but we can find a solution such that J is bounded (Simon, 2001) as follows:

$$J < \frac{1}{\gamma} \tag{5}$$

The constant γ to be used in establishing the bound can be determined from the following equations (Zhou, 1996):

$$AY + YA^* - YC^*CY + BB^* = 0 \tag{6}$$

$$Q(A - YC^*C) + (A - YC^*C)^*Q + C^*C = 0 \tag{7}$$

$$\gamma_{\min} = \frac{1}{\sqrt{1-\lambda_{\max}(YQ)}} \tag{8}$$

$$\gamma = \alpha\gamma_{\min} \qquad \alpha > 1 \tag{9}$$

We first calculate Y from the Riccati equation (6) and then calculate Q from the Lyapunov equation (7). Next, we set γ to be some number larger than $\gamma_{\min}$. The quantity $\gamma_{\min}$ is the smallest Eigen value of matrix YQ.

Once we have determined γ, we can reduce the H_∞ filtering problem to four equations (Simon, 2001):

$$L_k = (I - \gamma Q P_k + C^T V^{-1} C P_k)^{-1} \tag{10}$$

$$K_k = A P_k L_k C^T V^{-1} \tag{11}$$

$$P_{k+1} = A P_k L_k A^T + W \tag{12}$$

$$\hat{x}_{k+1} = A\hat{x} + Bu_k + K_k(y_k - C\hat{x}_k) \tag{13}$$

Where Q is a matrix that weighs the relative importance of each state, W is a matrix that weighs the importance of the process noise vector and V is a matrix the weighs the importance of the measurement noise vector. Fortunately, the filter gain matrix, K_k converges to a constant after only a few time steps, so we can compute this off-line and then just use the constant matrix in the state estimation equation (13).

Before demonstrating how H_∞ filtering works, we will introduce the commonly used minimum variance (Kalman) filter. This filter can be computed from the following three equations (Simon, 2001):

$$K_k = A P_k C^T (C P_k C^T + S_z)^{-1} \tag{14}$$

$$P_{k+1} = A P_k A^T + S_w - A P_k C^T S_v^{-1} C P_k A^T \tag{15}$$

$$\hat{x}_{k+1} = A\hat{x}_k + Bu_k + K_k(y_{k+1} - C\hat{x}_k) \tag{16}$$

where S_w is the covariance matrix of the process noise, S_v is the covariance matrix of the measurement noise and P is the variance of the estimation error. Again, K_k converges to a constant after only a few time steps.

We can see the similarities between the Kalman filter and the H_∞ filter from equations (10) through (13) and equations (14) through (16). Because of this, the Kalman filter is often referred to as the H_2 filter, as it attempts to minimize the 2-norm (variance) of the estimation error. In a similar sense, the H_∞ filter is named because it attempts to minimize the ∞-norm (maximum value) of the estimation error.

4. COMPARISON OF OPTIMAL FILTERING

To compare these two filtering techniques, we should first point out that the theory behind Kalman filtering is more mature than that of H_∞ filtering and as a result, Kalman filters are more widely used in practice.

The system simulated here is a simple navigational system consisting of two states: position and velocity. This system was chosen to allow an easy comparison of the results of both methods of state estimation. For the model, we assume a simple linear acceleration of 1 m/second^2 and can therefore write the equation for velocity as:

$$v_{k+1} = v_k + Tu_k + w_k \tag{17}$$

where v is the velocity, T is the sampling time period, u is the input (acceleration) and w is the velocity measurement noise. We can also write the equation for position as:

$$p_{k+1} = p_k + Tv_k + \frac{1}{2}T^2 u_k + z_k \tag{18}$$

where p is the position, and z is the position measurement noise. From equations (17) and (18), we can simulate the results from state estimation for the case where we have an accurate model of the noise and the case where we have an inaccurate model of the noise. For these simulations, we assume we are measuring both position and velocity.

Figure 1 below shows the velocity estimation error for the case where we have accurately modeled the noise. In this figure, we see that the Kalman filter provides better performance.

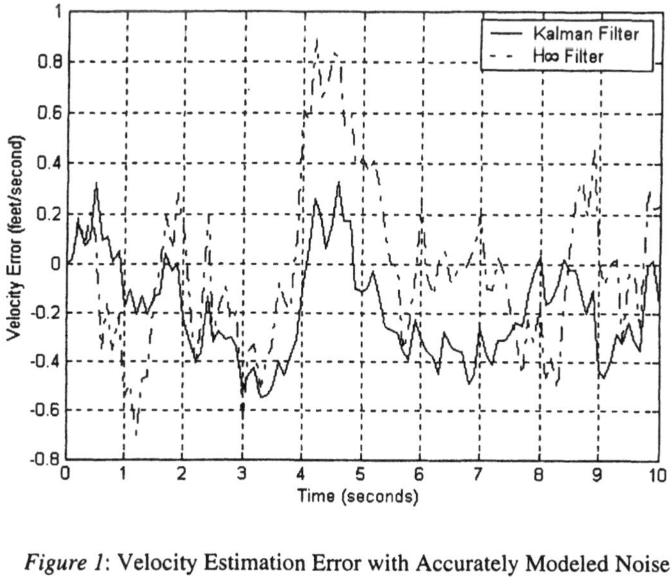

Figure 1: Velocity Estimation Error with Accurately Modeled Noise

Figure 2 below shows the position estimation error again for the case where we have accurately modeled the noise. In this figure, we see that both the H_∞ filter and the Kalman filter achieve almost the same performance. Looking at figures 3 and 4, we can see that the Kalman filter provides better overall performance than the H_∞ filter in the case where we have an accurate model of the noise.

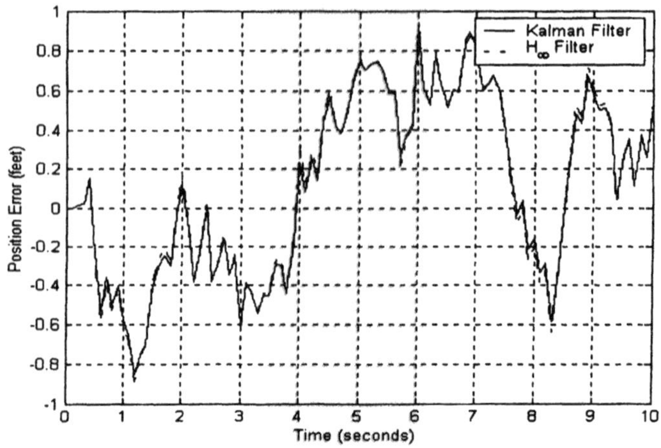

Figure 2: Position Estimation Error with Accurately Modeled Noise

Now that we have discussed the case where we have an accurate model of the noise, we can look at the case where we have an inaccurate model of the noise. Figure 3 below shows the velocity estimation error for this case. In this figure, we see that the H_∞ filter provides much better performance.

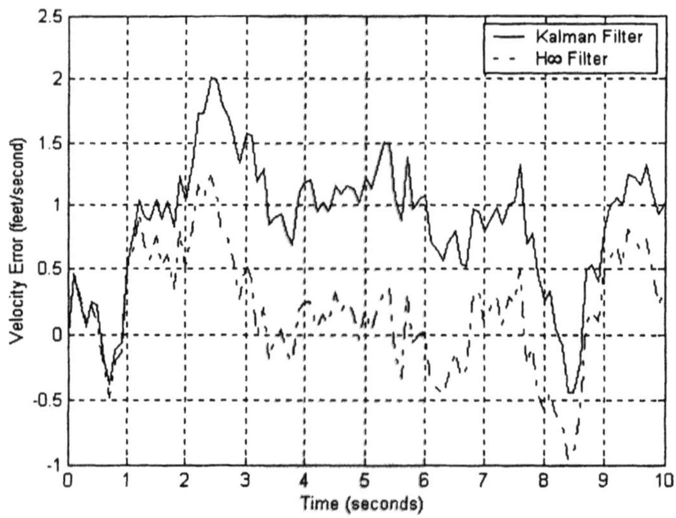

Figure 3: Velocity Estimation Error with Inaccurately Modeled Noise

Figure 4 below shows the position estimation error again for the case where we have not accurately modeled the noise. Once again, for the position estimate, the H_∞ filter and the Kalman filter achieve almost the same performance. From figures 3 and 4, we can see that the H_∞ filter provides better performance than the Kalman filter in the case where we have an inaccurate model of the noise.

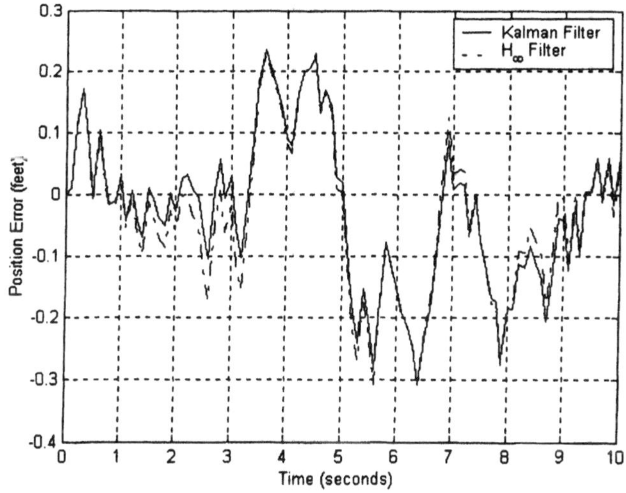

Figure 4: Position Estimation Error with Inaccurately Modeled Noise

As shown in figures 1 through 4, Kalman filters generally provide better performance when the system dynamics are fully understood and adequately modeled. On the other hand, H_∞ filters provide more robust estimates, as they are insensitive to uncertainties, but they generally require more tuning than Kalman filters (Simon, 2001).

5. CONCLUSIONS AND FUTURE WORK

As shown in section IV, the computer simulations indicate that the H_∞ based control systems have promise for applications in the adaptive optical systems. These estimation techniques can provide more robust wavefront estimates than conventional optimizations. An added benefit to the use of H_∞ techinques over standard techniques is that we can move a working system from one location to another without having to recalculate noise covariances for the filter equations.

Extensions to this work include implementing the system along with an H_∞ controller and comparing results from applying H_∞ and H_2 techniques to an adaptive optics system. Other research areas to be worked on include using a basis more conducive to the computational determination of the optical wavefront, such as a wavelet basis and applying system identification techniques to the calibration procedure.

This work was supported in part by Office of Naval Research grant number N00014-02-1-0022, by Xinetics, Inc. and by the North Carolina Space Grant Consortium.

6. REFERENCES

Michaud P, Kraft L, Liu MC. Press Release, January 7, 2002.

Hudgin, Richard. Optimal Wave-front Estimation. J. Opt. Soc. Am. 1977; 67: 378-82.

Irwan R, Lane RG. Analysis of Optimal Centroid Estimation Applied to Shack-Hartmann Sensing. Applied Optics 1999; 38: 6737-43.

Paschall RN, Anderson DJ. Linear Quadratic Gaussian Control of a Deformable Mirror Adaptive Optics System with Time-Delayed Measurements. Applied Optics 1993; 32: 6347–58.

Wallner EP. Optimal Wave front Correction Using Slope Measurements. J. Opt. Soc. Am. 1983; 73: 1771-76.

Ellerbroek BL, Van Loan C, Pitsianis NP, Plemmons RJ. Optimizing Closed-Loop Adaptive-Optics Performance with use of Multiple Control Bandwidths. J. Opt. Soc. Am. 1994; A11: 2871-86.

Ellerbroek BL, Rhoadarmer TA. Real-time Adaptive Optimization of Wave front Reconstruction Algorithms for Closed-Loop Adaptive-Optical Systems. SPIE Proc. 1998; 3353: 1174–85.

Hardy, John W. Adaptive Optics for Astronomical Telescopes, New York: Oxford University Press, 1998.

Tyson, Robert K. Introduction to Adaptive Optics, Bellingham, WA: SPIE Press, 2000.

Franklin GF, Powell JD, Workman ML. Digital Control of Dynamic Systems, 2nd ed. Reading, MA: Addison-Wesley, 1990.

Furber ME, Jordan D. Optimal Design of Wave front Sensors for Adaptive Optical Systems: Part 1, Controllability and Observability Analysis. Opt. Eng. 1997; 36: 1843–55.

Zhou K, Doyle JC, Glover K. Robust and Optimal Control. New Jersey: Prentice Hall, 1996.

Simon D. From Here to Infinity. Embedded Systems Programming. 2001; October: 2-9.

SMALL LINEAR ACTUATOR FOR SMALL ROBOTS

Hiroshi OKABE、 Naotake KAKIZAWA
Susumu SAKANO and Munekazu KANNO
College of Engineering, Nihon University
1 Nakagawara-Tokusada Tamuracho
Koriyama 963-8642 Japan

This paper presents the basic concept of the new small actuator, which is consisted of bimorph type piezoelectric element. This is an actuator for manipulating small and micro devices and micro organisms. The principle of the actuator and the static and dynamic characteristics of the actuator are shown in this paper .The actuator is driven using the principle of the hysteresis of the piezoelectric material.

Keywords: Actuator, Nonlinear characteristics, Piezoelectric element, Hysteresis of material, Micro actuator

1. INTRODUCTION

The application of robots under the endoscope such as the operation to the medical field is examined recently. By the operation under the endoscope, the low invasion operation becomes possible to the living body. And, the positioning of submicron or nanometer is carried out in the biomedical field such as the cell fusion, and various actuators are developed and are used. By the advantage which enables that the piezoelectric device is small and that it can produce large force, it is used as the actuator in medical treatment and in the biomedical field. The search about the high-precise positioning mechanism using rapid deformation of the layered type piezoelectric elements and the X-Y-Z positioning table using three poles cylindrical piezoelectric device have been carried out and the results of the studies were announced until now. The driven principle of the above mentioned mechanisms are the balance of inertia , friction and sliding between the driver and the driven body.. The motion of the mechanism is instable and it is impossible that the large force arises.

We have studied the small actuator using the bimorphic piezoelectric device and the rotary type actuator and the linear type actuator have been developed. However, it was difficult that the actuators could not produce the large force, because the phenomenon of the contact was used and they could not produce the large force.

The actuator using the new driving method is discovered recently on the basis of the above mentioned studies. The actuator is driven using the hysteresis of the material.

The driving principle of the new actuator is explained and the result of the trial manufacture as the linear actuator is shown in this paper. The driving principle of the actuator is completely new and the actuator is driven by the non-contacting force. The actuator does not use friction and sliding between the driver and the driven body like the actuator using the conventional piezoelectric devices. The new actuator is applied to the actuator of the small robots and will be applied to the precise positioning table and the medical treatment.

2. DRIVING PRINCIPLE OF ACTUATOR

2.1 Driving principle

The driving principle of the actuator is shown in Figure 1. The bimorphic piezoelectric element is supported by the driven body. The piezoelectric element is like a cantilever beam. When the cyclic drive voltage is loaded to the piezoelectric element, the piezoelectric element is vibrated and the driven body moves to right direction or to left direction. The driven force of the piezoelectric element is produced from the nonlinear characteristic of the material hysteresis. The driven body is moved by the nonlinear deformation of the bimorphic piezoelectric element.

The body is moved by the noncontacting force

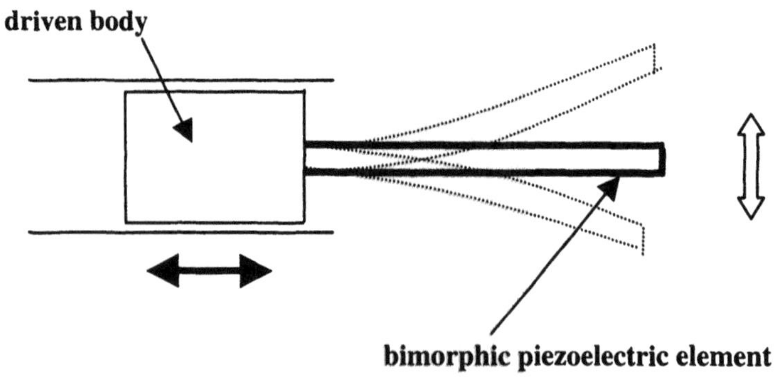

Figure 1 Driving principle of actuator

2.2 Estimation of driving force

The assumptions for requiring the driving force are as follows:

(1) the nonlinear characteristics of the bimorphic piezoelectric element is approximated some straight lines.

(2) the displacement characteristics of the element to the driven voltage is shown by the cantilever beam which receives the uniform load.

The nonlinear characteristics of the bimorphic piezoelectric element with the driven voltage is shown as Figure 2. The bimorphic piezoelectric element has the hysteresis characteristic to the driven voltage. The broken line in the figure is the actual characteristics and this characteristics are approximated by two straight lines. The lines are shown as the solid lines. These lines are shown as the next two equations.

$$Y_1(V) = aV - b \tag{1}$$

$$Y_2(V) = aV + b \tag{2}$$

Where, V is the driving voltage and a, b are fixed numbers.

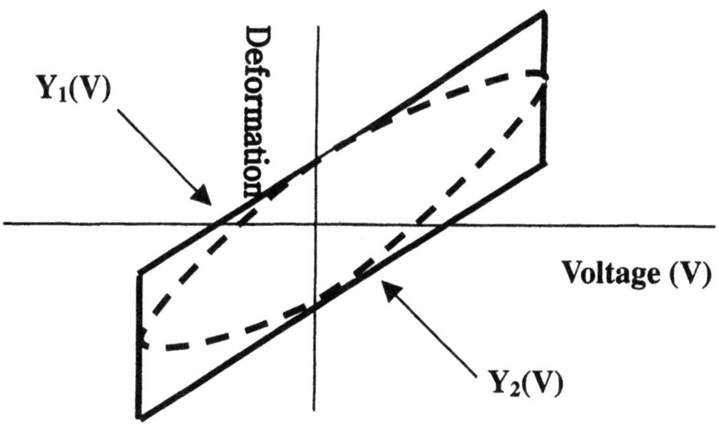

Figure 2 Nonlinear characteristics of piezoelectric element

The piezoelectric characteristics is expressed as the deformation of the cantilever beam which receives the uniform load . The load per unit length of the uniform load is shown as w. The relationship between the deformation and the one load w is expressed as Figure 3. The driving force to the driven body by this one load w is solved from the relationship between the deformation of the cantilever beam. The force is expressed as the next equation.

$$F_i(x, V) = w\, Y_i\,(V) \sin\, \alpha \tag{3}$$

$$\alpha = w\ Y_i\,(V)\, x^2 / 2EI \tag{4}$$

Where, E is young modulus and I is inertia moment of cross section of the cantilever beam.

The total driving force to the driven body is required by integrating from the root of the cantilever beam to the tip of the beam.

$$F_i(V) = \{w\ Y_i(V)\ \}^2\, x^2 / 2EI \tag{5}$$

$$F_1(V) = \int_0^l \frac{\{wY_1(V)\}^2}{2EI} x^2 dx \tag{6}$$

$$F_2(V) = \int_0^l \frac{\{wY_2(V)\}^2}{2EI} x^2 dx \tag{7}$$

$$F_1(V) = \frac{\{wY_1(V)\}^2}{2EI} l^3 \tag{8}$$

$$F_2(V) = \frac{\{wY_2(V)\}^2}{2EI} l^3 \tag{9}$$

$Y_1(V)$ is the upper straight line and $Y2(V)$ is the lower straight line in Figure 2. The force ($F_1(V)$ - $F_2(V)$) works to the driven body and the driven body moves by this force.

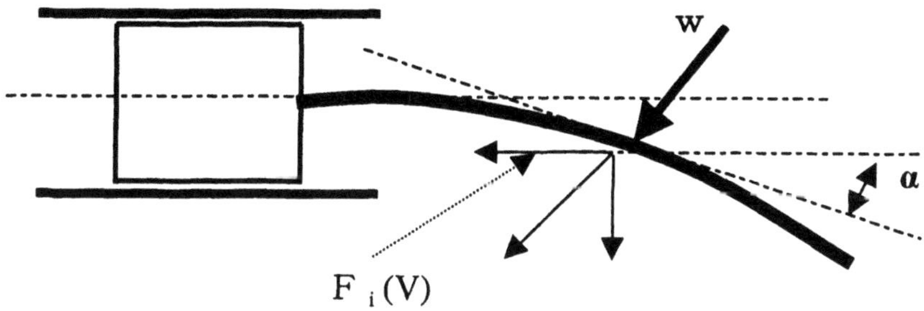

Figure 3 Deformation and force by load w

$$F_1(V) - F_2(V) = \frac{2w^2 l^3}{3EI} ab\, v^{*2} \tag{10}$$

Where, v^{*2} is the maximum driving voltage to the piezoelectric element.

The uniform load w per unit length of the cantilever beam is obtained from the relationship between the equation (10) and the deformation of the piezoelectric beam which receives the driving voltage. The load w is obtained as next equation.

$$w = \frac{8EI}{l^4} \tag{11}$$

Finally, the driving force to the driven body is expressed as the next equation.

$$F_1 - F_2 = \frac{128EI}{3l^5} ab v^{*2} \tag{12}$$

The driving force is estimated using the concrete values as shown in Table 1. The generated force is different by Young modulus of the piezoelectric material. Table 2 shows the generated force.

Table 1 Parameters in estimating driving force

Parameter	Value
I	0.216 mm^4
a	$6.67 * 10^{-3}$
b	0.1 mm/V
l	45 mm
V	60 V

Table 2 Generated force

Young modulus	Generated force
$5 * 10^3$ kg/mm^2	0.000196 N
$5 * 10^4$	0.00196
$5 * 10^5$	0.0196

The generated force of the 1cycle voltage (from +60 V to - 60V) is changed from 0.0196 N (2g) to 0.00196 N (0.02g) by the) Young modulus of the piezoelectric material.

The driving force is generated using nonlinear characteristics and hysteresis property of the piezoelectric element and the driven body is moved by this force. The above mentioned principle is applied to the actuator.

3. EXPERIMENTS

3.1 Experiment system

The experimental equipment is shown in Figure 4. The bimorphic piezoelectric element is supported in the moving body in the condition of the cantilever beam. The moving body is installed on the acrylic board and the acrylic board moves on the metal. The copper wire for power collection is placed on the acrylic board and the collected electric power is transmitted to the piezoelectric element.

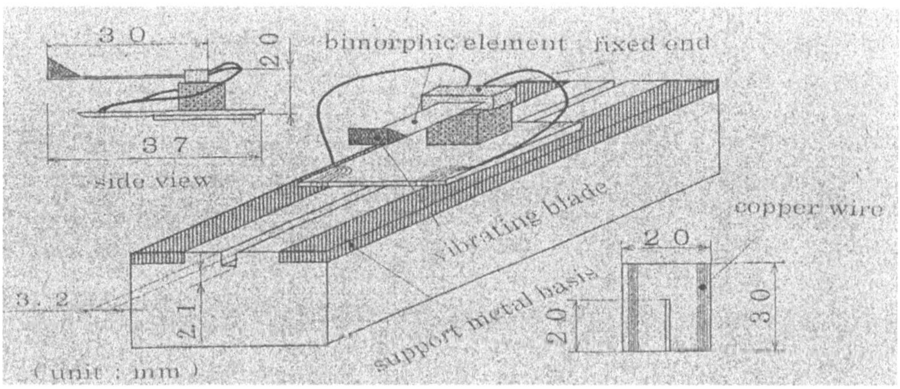

Figure 4 Experimental equipment

3.2 Experimental results

The experimental results are shown in Figure 5 and Figure 6. Figure 5 shows the frequency characteristics of the moving body. The moving body gets forward and gets backward according to the changes of the frequency. The forward maximum speed is obtained near 340 Hz and the backward maximum speed is obtained near 1650 Hz. Figure 6 shows the relationship between the electric charged voltage to the piezoelectric element and the speed of the moving body. The speed of the moving body becomes constant over 60 V of the charged voltage.

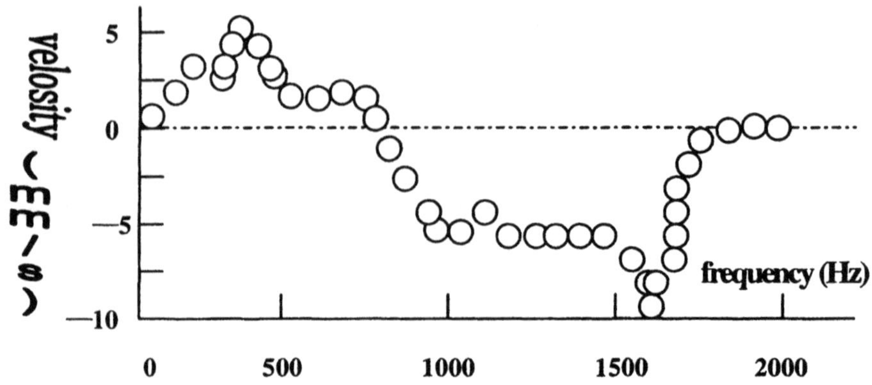

Figure 5 Frequency characteristics of the actuator

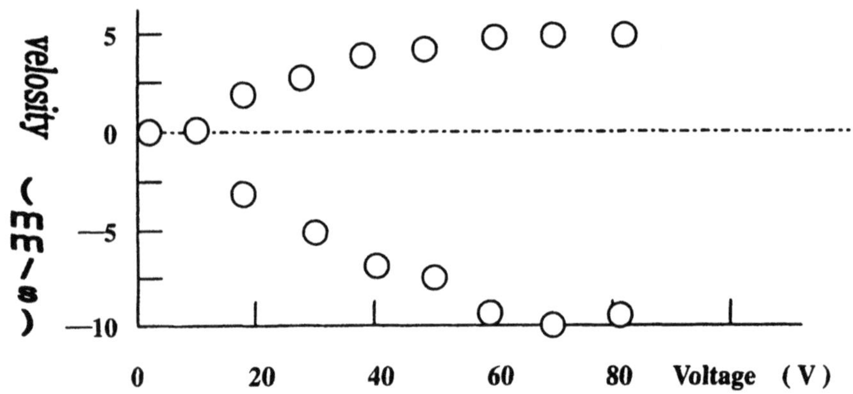

Figure 6 Moving speed of moving body

4. CONCLUSIONS

The new driving method using the piezoelectric device was presented. The drive principle was shown and the new method is applied to the actuator. The driving characteristics of the actuator are obtained by the experiments. The driving method is the non-contacting drive unlike the conventional contacting type drive using the piezoelectric device.

5. REFERENCES

1. E. Shamoto, H. Shin, T. Moriwaki, Development of precision feed mechanism by applying principle of walking drive, JSPE, 59−2, 1993, pp. 317-322.
2. T. Honda, T. Sakashita, K. Narahashi , and J. Yamasaki, Swimming properties of a bending-type magnetic micro-mechanism, J. of Japan Applied Magnetic Society, 25-4-2, 2001, pp. 1175-1178.
3. T. Higuchi, M. Watanabe, K. Kudo, Precise positioner utilizing rapid deformation of a piezoelectric element, JSPE, 54-11, 1988, pp. 2107-2112.
4. T. Honda, K. Ishiyama, K. Arai, Micro swimming mechanism propelled by external magnetic fields, IEEE. Trans. Magn. 32-2, 1996, p. 5085.
5. T. Niino, S. Segawa, T. Higuchi, High-power and high-efficiency electrostatic actuator, Proc. IEEE Work-shop on MEMS, 1993, p. 236.
6. A. Sadamoto, et al., Wireless micromachine for in-pipe visual inspection and possibility of biomedical applications, Proc. 32nd International symposium on Robotics, 2001, p. 433.
7. S. Hata, Y. Yamada, J. Ichihara and A. Shimokohbe, A micro lens actuator for optical flying head, Tech. Digest MEMS 2002, 2002, p. 507
8. M. Ookoshi, S. Sakano, Microstep X-Y- θ table using three-pole piezoelectric tube actuator, J. JSME, 62-596, 1996, pp. 1392-1396.

A NEW ELECTROMAGNETIC ACTUATION SYSTEM ON AN INDUSTRIAL SEWING MACHINE WITH ON-LINE EFFICIENCY MONITORING

Luís F. Silva[(1)], Mário Lima[(1)], Fernando N. Ferreira[(2)], Ana M. Rocha[(2)], Helder Carvalho[(2)], Carlos Couto[(3)]

[(1)] Dep. of Mech. Eng. [(2)] Dep. of Textile Eng. [(3)] Dep. of Ind. Electronics, School of Eng. University of Minho, Campus de Azurém, 4800-058 Guimarães, Portugal Tel: 351 253 510226, Fax: 351 253 516007, mlima@dem.uminho.pt

This paper briefly reviews the study to evaluate the standard presser foot performance and a new actuation set-up to avoid the bouncing and the lack of fabric control. The compression force and displacement waveforms are also presented and widely discussed, as well as the seam's quality analysis. A spectral (Fast Fourier Transform and Harmonic Distortion) analysis on the obtained waveforms, as well as an Admissible Displacement Limits (ADL) analysis are also described as important techniques to be used to supervise the presser foot force and to monitor the feeding efficiency. ADL have proved to be a better method. Further research should be undertaken in this area, in order to achieve, in a near future, a closed loop control of the presser foot.

Keywords: Overlock sewing machine, Fabrics feeding system, Presser foot, Proportional force solenoid, Feeding efficiency monitoring.

1. INTRODUCTION

The feeding system of an overlock sewing machine is made up by a presser foot, a throat plate and a feed dog. At high sewing speeds the presser foot "*bounces*" and loses contact with the fabric plies resulting on an irregular stitch. To study and evaluate its dynamic behaviour, a "sewability" tester was first developed by Rocha, *et al.* (Rocha, 1996), and later improved by Carvalho and Ferreira (Carvalho, 1996), and Carvalho, *et al.* (Carvalho, 1997 & 2000). An industrial overlock sewing machine was instrumented with miniature piezoelectric force transducers on the presser foot and needle bars, encoders and semiconductor strain gauge transducers on the threads' path. A LVDT for real-time monitoring the movement of the presser foot bar led to further understanding the system dynamics. Figure 1 shows typical presser foot bar displacement waveforms obtained with two plies of a jersey fabric, using a presser foot force of 20N, a uniform feed and a stitch density of 4 stitches per cm. The beginning of the stitch cycle (0 degrees) is set when the needle is at its lowest position, after fabric penetration. The compression force waveforms obtained with two plies of the same jersey fabric are presented in figure 2 and show, during the feeding period (between 80 and 260 degrees), wide variations with the increase of machine speed (in stitches per minute, spm). At 260 degrees the final peak forces indicate the instant when the presser foot re-establishes contact with the throat plate.

These force variations suggest that the spring forces are not adequate to guarantee an efficient contact between the presser foot and the fabrics. Both contact losses highlighted in figure 1 (at the topmost position of the feed dog above the throat plate level – peak 1 – and soon after the feed dog has been lowered beneath the same level – peak 2 –) occur in all fabrics sewn when the compression forces show a tendency to decrease, during and after the feeding process respectively. Therefore, a precise and accurate control of the fabric plies is essential to reduce, or even eliminate, the loss of feeding efficiency that reduces the quality of the seams (Silva, 1999) & (Silva, 1999).

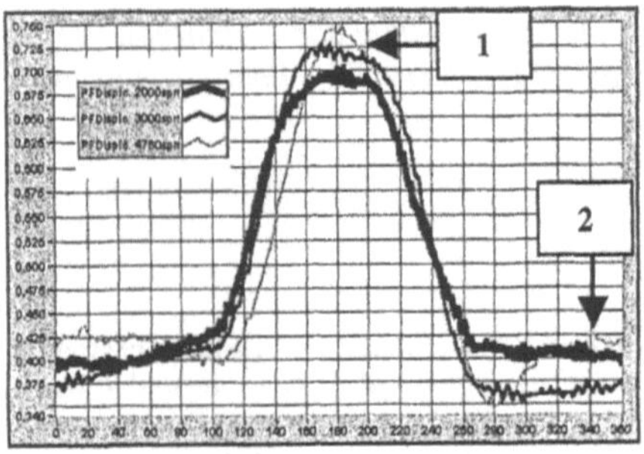

Figure 1 – Horizontal axis: stitch cycle (degrees); Vertical axis: presser bar displacement (mm)

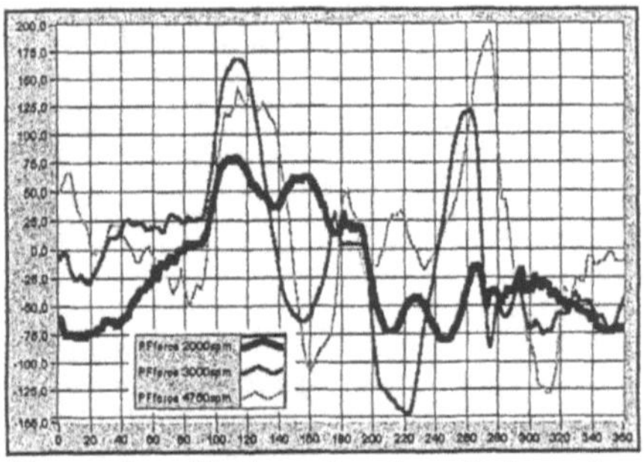

Figure 2 – Horizontal axis: stitch cycle (degrees); Vertical axis: presser bar compression force, relative values (gf)

2. EXPERIMENTAL SET-UP AND PROCEDURE

A *proportional force solenoid* was found to be the most suitable in order to achieve the desired presser foot dynamic behaviour (Silva, 1999). This actuator replaced the standard helical spring, without modifying the arrangement of the force and displacement transducers already included in the previous phase –figure 3.

Experiments were carried out at different machine speeds. Two plies of three different knit fabrics were used (jersey, interlock and rib 1×1) and two operating conditions were tested: (1) *With* and (2) *Without* the helical spring. The other machine settings were kept unchanged throughout the experiments.

The actuator, presenting a constant force characteristic (independent of the displacement) will be working in a "continuous" mode, emulating a spring with programmable reaction force. The current to drive the solenoid, to apply an additional or total force, was correlated with the wide force variations observed for all tested fabrics.

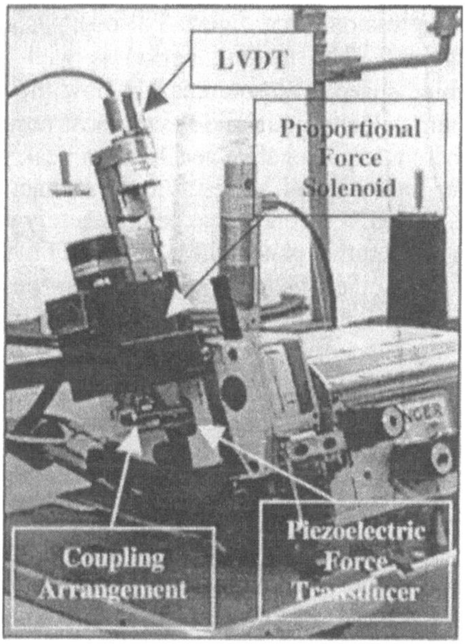

Figure 3 – The presser foot actuation set-up with the proportional force solenoid attached to the presser bar

3. RESULTS AND DISCUSSION

3.1 Using the Solenoid and the Helical Spring

Figures 4 and 5 show, respectively, the presser foot bar displacement and compression force waveforms obtained at 2000, 3000 and 4700spm with the jersey fabric. It is clear that the two contact losses observed in figure 1 (peak 1 and 2) have completely disappeared and the plies are under control, not only in the feeding period but also after the feed dog has been lowered beneath the throat plate level.

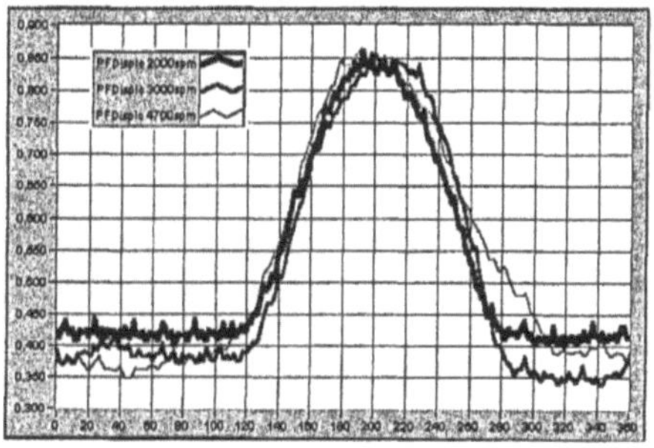

Figure 4 – Horizontal axis: stitch cycle (degrees); Vertical axis: presser bar displacement (mm)

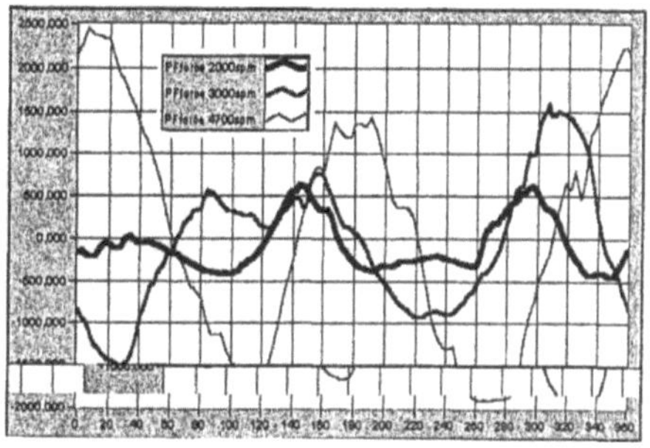

Figure 5 – Horizontal axis: stitch cycle (degrees); Vertical axis: presser bar compression force, relative values (gf)

In terms of compression force, figure 5 also shows a different waveform when compared with figure 2. The feeding period is well defined, showing a force increase with machine speed. The solenoid has now the presser foot under control between its maximum velocity point and the topmost position of the feed dog above the throat plate, where a large negative acceleration peak was previously observed – see (Silva, 1999) for further details –, resulting on contact losses detected in figure 1 (peak 1). After this period, a second force peak is surely responsible for holding the fabrics firmly against the throat plate, where a second higher possibility for a contact loss may occur, as another negative acceleration peak was observed, which corresponds to peak 2 in figure 1. This important peak also increases with machine speed.

3.2 Using the Solenoid

Once again, no contact losses have been observed for the jersey fabric – see figure 6. Figure 7 shows the corresponding presser foot bar compression force waveforms obtained for the same machine speeds. The feeding period is better defined and clearly without the wide force variations shown in figure 2.

At 4700spm (figure 6), the decrease of the maximum displacement and the fabrics' compression occurred after feeding, was probably caused by applying a force slightly higher than the amount required for controlling the jersey plies; this same effect has also been observed for the other two fabrics. To test this hypothesis, another set of experiments was carried out for the this same fabric varying the force applied to the presser bar, between 16.70N and 23.46N, which corresponds to a variation of the drive currents between 0.403A and 0.503A. The upper limit equals the force applied by the solenoid at this same speed. The obtained presser foot bar compression force and displacement waveforms, while testing three different drive currents at 4700spm, are shown in figures 8 and 9, respectively.

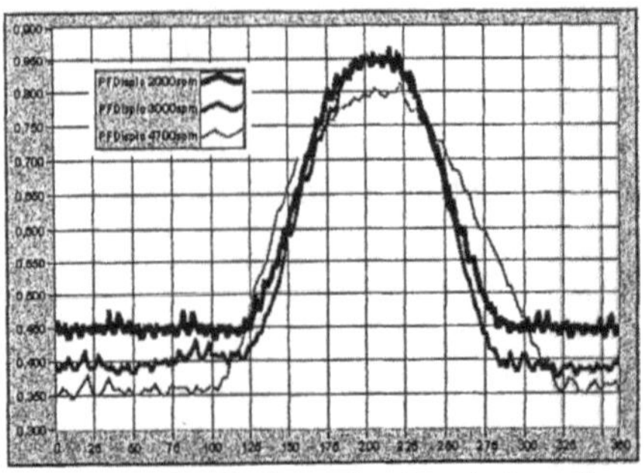

Figure 6 – Horizontal axis: stitch cycle (degrees); Vertical axis: presser bar displacement (mm)

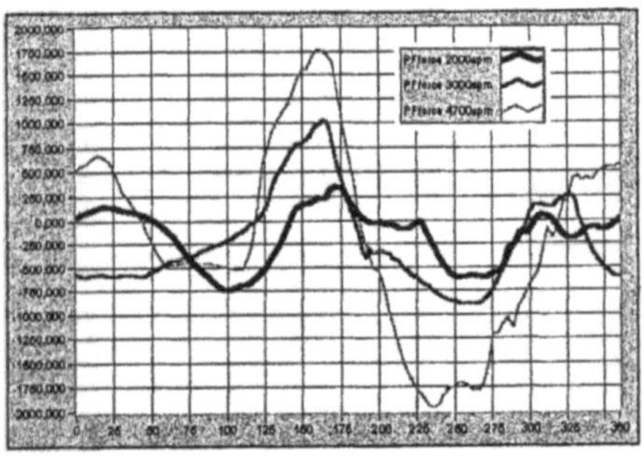

Figure 7 – Horizontal axis: stitch cycle (degrees); Vertical axis: presser bar compression force, relative values (gf)

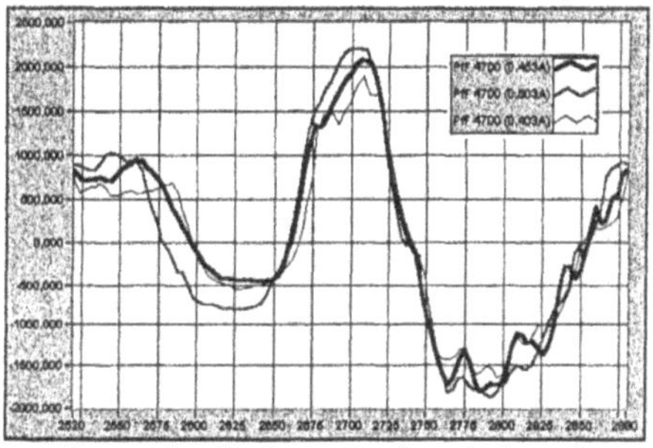

Figure 8 – Horizontal axis: stitch cycle (degrees); Vertical axis: presser bar compression force, relative values (gf)

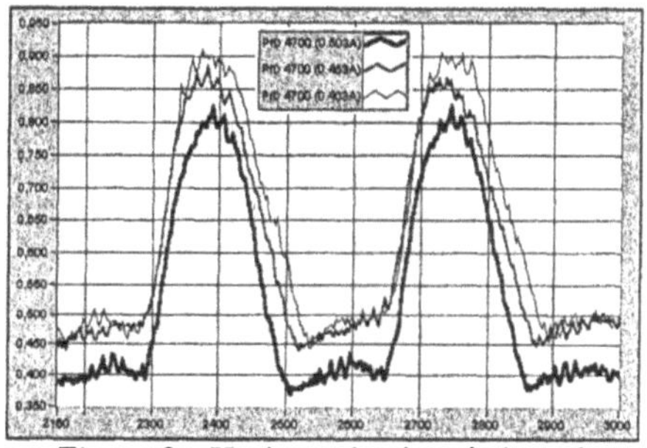

Figure 9 – Horizontal axis: stitch cycle (degrees); Vertical axis: presser bar displacement (mm)

When the force applied to the presser bar decreases, other force variations tend to appear during and after the fabrics feeding. Under these circumstances, the presser foot is unable to guarantee an effective control of the fabric plies, and an increase of the presser foot bar maximum displacement has been observed. When the needle penetrates and withdraws from the fabrics to form the stitch (after the feed dog has been lowered beneath the throat plate), a second displacement also occurs. For the lowest forces, these displacements are even greater than the ones observed in figure 4, using the solenoid and the helical spring, and in figure 6, with only the proportional force solenoid. The waveforms obtained for the other knitted materials (not presented here) revealed exactly the same: the decrease of the forces affects the performance of the presser foot and suggests that the force initially applied provides an effective control of the fabrics.

3.3 Seams Quality

An efficient feeding system should be able to guarantee the same stitch density, independently of the fabrics or speeds involved. Ideally, the obtained stitch density should always match the one defined with nominal machine settings. The following tables summarise the mean number of stitches per cm (spcm) counted over several seams measurements of 5 cm long for all tested samples.

Using the helical spring on the presser foot bar, the differences between the actual obtained stitch density and the nominal ones increase with machine speed. For the jersey fabric, this loss of efficiency can be explained according to the contact losses observed in figure 1 (peaks 1 and 2). Due to this "bouncing", the presser foot is unable to control the fabrics and the stitch length tends to increase with machine speed, decreasing, therefore, the obtained stitch density – see table 1. The interlock and rib 1×1 fabrics show, however, a different sewing behaviour: the stitch densities increase with machine speed, being the maximum difference variations (of +5.5% and +11.5%, respectively) found at 4700spm.

On the other hand, these differences have decreased using the new actuation system, with or without the helical spring. For the jersey fabric (table 1) it can be observed that there is a tendency in both cases to increase the control of fabrics feeding; however, the minimum stitch density differences have been observed with just the solenoid for all tested speeds. These differences have also decreased for the interlock and rib fabrics using only the solenoid on the presser foot bar. Tables 3 and

4 indicate, for 0.502A, the density variations obtained at 4700spm (-0.5% and +0.75%, respectively), which represents a huge reduction when compared with the differences already mentioned.

Table 1 - Nominal and actual stitch densities; two plies of a jersey fabric

System type	Machine speed (spm)	Nominal stitch density (spcm)	Actual stitch density (spcm)	Variation (%)
Helical spring	2000	4	3.80	-5.0
	3000		3.75	-6.25
	4700		3.65	-8.75
Solenoid and spring	2000	4	3.85	-3.75
	3000		3.95	-1.25
	4700		4.03	+0.75
Solenoid	2000	4	3.98	-0.5
	3000		4.03	+0.75
	4700		4.05	+1.25

Table 2 - Stitch densities; two plies of a jersey fabric changing the drive current

Current (A)	Machine speed (spm)	Nominal stitch density (spcm)	Actual stitch density (spcm)	Variation (%)
0.503	4700	4	4	0
0.483			3.9	-2.5
0.453			3.87	-3.25
0.403			3.84	-4.0

The obtained stitch densities varying the drive currents are presented in tables 2, 3 and 4. A decrement of the force applied to the presser foot bar led to the increment of the density differences. These differences show the same tendency as observed before for each fabric using the standard presser foot and are now much smaller. It must be noticed that a null density variation has also been found for the jersey fabric with these set of experiments (see also table 2). These results show that the calculated force (current) gives clearly better results.

Table 3 - Stitch densities; two plies of an interlock fabric changing the drive current

Current (A)	Machine speed (spm)	Nominal stitch density (spcm)	Actual stitch density (spcm)	Variation (%)
0.502	4700	4	3.98	-0.5
0.453			3.98	-0.5
0.403			4.17	+4.25

Table 4 - Stitch densities; two plies of a rib 1×1 fabric changing the drive current

Current (A)	Machine speed (spm)	Nominal stitch density (spcm)	Actual stitch density (spcm)	Variation (%)
0.502	4700	4	4.03	+0.75
0.453			4.08	+2.0
0.403			4.16	+4.0

4. THE ON-LINE MONITORING OF FEEDING EFFICIENCY

To provide an alternative perspective to characterise the behaviour of the standard presser foot assembly and the new actuation set-up, a Fast Fourier Transform (FFT) analysis was carried out on the obtained waveforms. This powerful signal analysis tool, included in the software package and developed according to project requirements – see also (Carvalho, 2000) for details and other FFT-related application examples –, was used to translate into the frequency-domain the acquired time-domain signals.

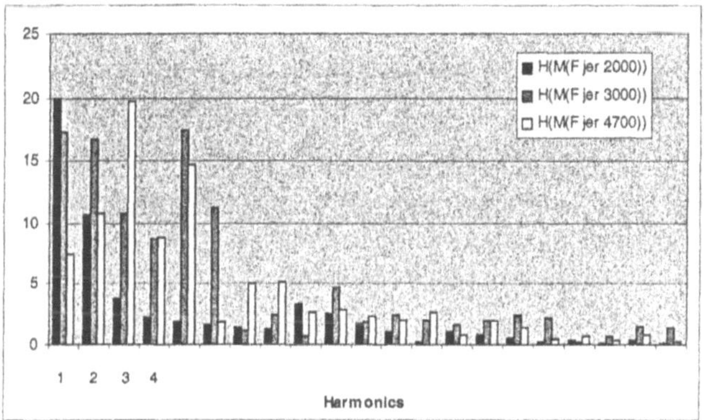

Figure 10 – Compression force harmonics obtained with the helical spring on the presser foot bar

Comparing the obtained displacement spectrums only slight amplitude changes occurred at higher frequency components. Nevertheless, completely different spectrums were observed after processing the presser foot bar compression force waveforms – see figures 10 and 11 for the jersey fabric. The standard spring actuation originates a more spread spectrum harmonics than the solenoid operated one, where it is clear a concentration in the low frequency range. These patterns stand in accordance, respectively, with the wide and quick force variations observed with the helical spring on the presser foot bar and the waveforms acquired using the solenoid, where only two force peaks were detected in the stitch cycle.

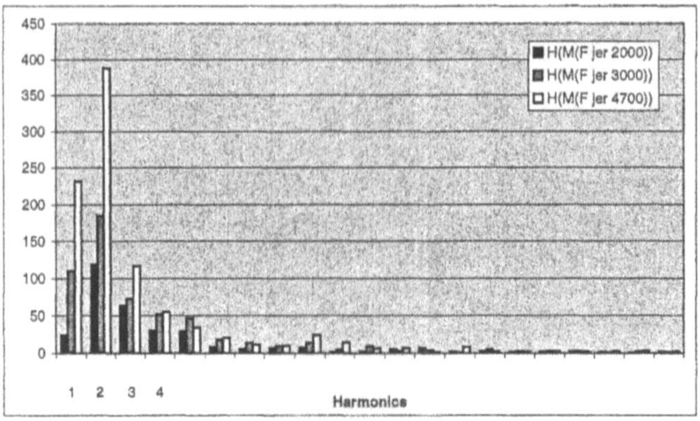

Figure 11 – Compression force harmonics obtained with the actuation set-up presented in figure 3

To develop an on-line monitoring system to supervise the presser foot force and to automatically detect the loss of feeding efficiency, a Total Harmonic Distortion (THD) analysis was established computing the previous results by using the following equation, where A_1 and A_i correspond to the harmonic amplitudes 1, i:

$$THD = \frac{\sqrt{\sum_{i=2}^{n} A_i^2}}{A_1} \qquad (1)$$

This equation was used to determine the THD of the compression force waveforms obtained with the helical spring, while the THD of the other waveforms were determined rearranging the previous equation and considering the amplitude of the double fundamental frequency harmonic as reference, instead of A_1. (Figure 11 displays this important harmonic component, also observed for the other tested fabrics.) The distortion for both cases is presented in figure 12. As expected, the distortion tends to increase with machine speed using the standard presser foot, due to the force variations already mentioned, while the best results were achieved using the actuation set-up of figure 3 with a clear increment of low harmonics.

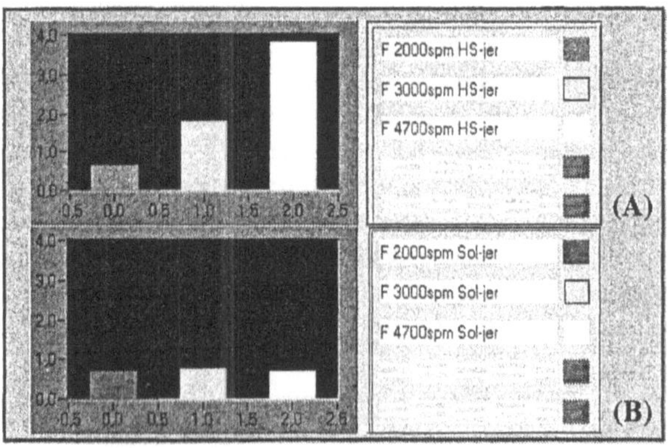

Figure 12 – Harmonic distortion on the obtained waveforms: (A) Using the standard helical spring (HS), and (B) Using the proportional force solenoid (Sol) as figure 3

The final feeding efficiency monitoring approach came after this spectral analysis. It was developed using the LabView graphical programming language, with specifically designed routines to automatically process, extract and display relevant features, enabling a complete analysis of the displacement and compression force waveforms and the evaluation of feeding efficiency, as a function of the imposed seams quality. As reported earlier, decreasing the force applied by the solenoid, using the set- -up in figure 3 and two plies of the same jersey fabric, the presser foot bar displacement shows a tendency to increase (see figure 13) and other force variations tend to appear, during and after feeding. Although the THD analysis has proved to be useful to characterise and identify the spring and solenoid actuation, in the latter case, according to the stitch density variations, the distortion changes are quite smaller, being extremely difficult to establish a distortion range to define the quality of the seams (see figure 14). Therefore, other investigations were directed to derivc an adequate method to properly monitor, under any circumstances, the fabrics feeding efficiency. Figure 13 illustrates the Admissible Displacement Limits (ADL), surrounding the 0% density variation, which establishes the upper and lower presser foot bar displacement limits according to the control values. If any maximum versus second displacement coordinate is plotted outside the admissible area, a warning display alerts the operator that the quality of the produced seam will not be the expected one. This method has been used successfully to on-line, and off-line, monitor the feeding efficiency.

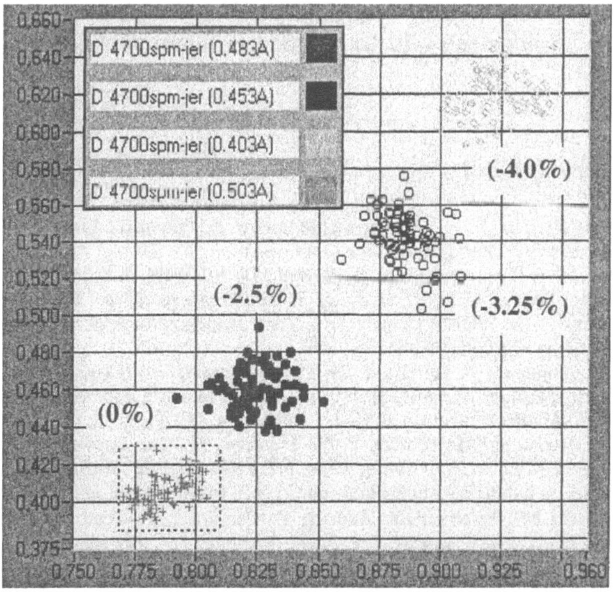

Figure 13 – Maximum versus second (after 260 degrees) displacements (mm) and stitch density variations obtained decreasing the force applied by the solenoid, according to figure 3

5. CONCLUSIONS

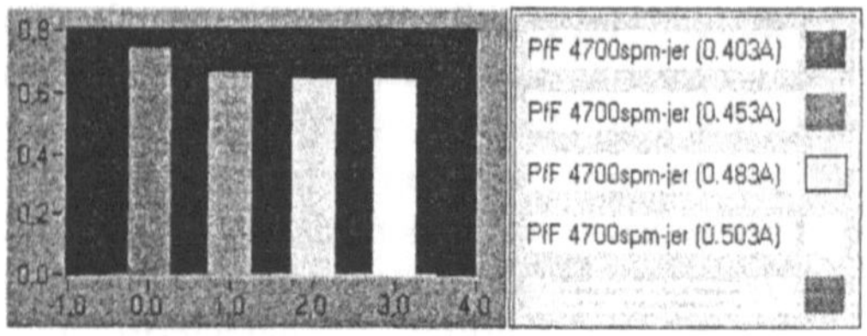

Figure 14 – Harmonic distortion on the obtained compression force waveforms with the solenoid attached as in figure 3

A new actuation system based on a proportional force solenoid was used replacing the standard spring actuated presser foot bar, aimed at avoiding the presser foot bouncing and lack of fabric control, especially at high speeds. This solution has shown clearly better results regarding the defects on the fabric plies and on the produced seams. THD (Total Harmonic Distortion) and ADL (Admissible Displacement Limits) were presented and discussed as methods for on-line monitoring the feeding efficiency. ADL have, so far, proved to be a better method.

Finally, it must be emphasised that the study reported in this paper was carried out until the end of 2001. Further research has been undertaken at the Laboratory of Process Research (University of Minho) and two presser foot (open and closed-loop) control schemes have already been implemented on the studied sewing machine. These recent developments will be reported in future publications (Silva, 2003) & (Silva, 2003).

6. REFERENCES

1. Rocha, A. M., Ferreira, F. N., Araújo, M., Monteiro, J., Couto, C., Lima, M. F., "*Mechatronics in Apparel: Control, Management and Innovation on the Sewing Process*", Proceedings of the Mechatronics'96 Conference, Vol. **II**, pp. 109-114, University of Minho, Guimarães, 1996.
2. Carvalho, M., Ferreira, F. N., "*Study of Thread Tensions in an Overlock Sewing Machine*", Proceedings of the ViCAM Conference – Vision and Control Aspects of Mechatronics, pp. 223-226. University of Minho, Guimarães, 1996.
3. Carvalho, H., Monteiro, J., Ferreira, F. N., "*Measurements and Feature Extraction in High-Speed Sewing*", Proceedings of the IEEE: ISIE'97 – International Symposium on Industrial Electronics, Vol. **3**, pp. 961-966. University of Minho, Guimarães, 1997.
4. Carvalho, H., Rocha, A. M., Ferreira, F. N., Monteiro, J., "*The Development of Support Tools for High-Speed Sewing Machine Setting, Monitoring and Control*", ISIAC'2000 – Third International Symposium on Intelligent Automation and Control, Maui, Hawaii, USA, June, 2000.
5. Silva, L. F., Lima, M., Ferreira, F. N., Andrade, D., Ferreira, L. P., Couto, C., "*Mechatronic Approach to the Feeding System of an Overlock Sewing Machine*". Mechatronics Journal, **9**, 817-842, 1999.
6. Silva, L. F., Lima, M., Ferreira, F. N., Coelho, J., Couto, C., "*Development of a Mechatronic Controlled Actuation on the Presser Foot of an Overlock Sewing Machine*", Proceedings of the 44th International Scientific Colloquium, Vol. **2**, pp. 116-121. Technical University of Ilmenau, Germany, 1999.
7. Silva, L. F., Lima, M., Carvalho, H., Rocha, A. M., Ferreira, F. N., Monteiro, J., Couto, C., "*Actuation, Monitoring and Control of Fabrics Feeding During High-Speed Sewing*", to be published in the Proceedings of the International Conference on Mechatronics – ICOM'2003, University of Loughborough, UK.
8. Silva, L. F., Lima, M., Carvalho, H., Rocha, A. M., Ferreira, F. N., Monteiro, J., Couto, C., "*Actuation, Monitoring and Closed-Loop Control of Sewing Machine Presser Foot*", to be considered for publication in the journal Transactions of the Institute of Measurement & Control, 2003.

PART FOUR
MANUFACTURING SYSTEMS INTEGRATION

UNDERSTANDING DATA IN INDUSTRIAL ENVIRONMENTS

Lopes, E. R., Mourinho, N. G., Castela, N., Guerra, A.
Escola Superior de Tecnologia do Instituto Politécnico de Castelo Branco
eurico@est.ipcb.pt

Industrial data is codified most of the times from different human languages and is supported in different databases systems and applications for industrial environments. Information retrieval, and knowledge can be done, if we have a human point of abstraction of data (Laudon, 1999). According to Fayyad (Fayyad et al, 1996), an understanding of the domain, extended database must exist with meaningful information. Using this assumption, we create an extended database, with two types of information: human knowledge of the fields and what kind of technique must be used to access data. With an easy human interface, we can model an accurate document system for industrial environments.

Industrial Data, Databases, Metaqueries, Metadata, Human Knowledge

1. INTRODUCTION

Lots of valuable information hidden in industrial environments is barely exploited, since we need abstract and high-level information that is tailored to the user's needs (Staud et al, 1998). Data collections from which to obtain this information have chaotic structures, different human languages, often erroneous, and it is only partial integrated. This data is result from most of production management systems, as MRPII, ERP even from CIM, FMS and also CAD/CAM/CAE systems (Sargent, 1997). Human perception is a crucial component for an effective discovery system (Shen et al, 1996). In order to find an understandable tuning data, we need a customizing tool that can perform necessary transformations automatically.

Although the data warehouse (English, 1999), already provides a consolidate and homogenous data collection as input from analysis tasks we have to deal with certain tasks and algorithm specific data preparation steps which includes a reasonable understanding of the industrial environment.
Human perception (Shen et al, 1996) is necessary for generating and selecting the most promising hypotheses in a practical manner. Metadata (Staud et al, 1998) plays also, an important role for the integration task, both at the design and run time, since it gives background knowledge; knowledge about data selection and transformations needed to properly feed outside systems and application methodology that gives criteria in which way to solve problems. Afterwards we create a logical documented

system, which help upgrades, finding information, and a practical documentation to avoid duplication of data and also to reduce "islands" of automation (Hannus, 1998).

This paper is organized as follows: first, a model to customize human words and mapping them to the databases systems, second the integration process of the model, and finally some results obtained in practical experiments.

2. MATHEMATICAL MODEL

For the purposed system, what we need is to modeling a human language representation that characterizes the micro-world of that environment.

This is done using the database structure definition (meta data) for a particular data base management system. We need a very high degree of portability, since what we want is a human representation of the data, which will be used for other retrieval or data mining techniques.

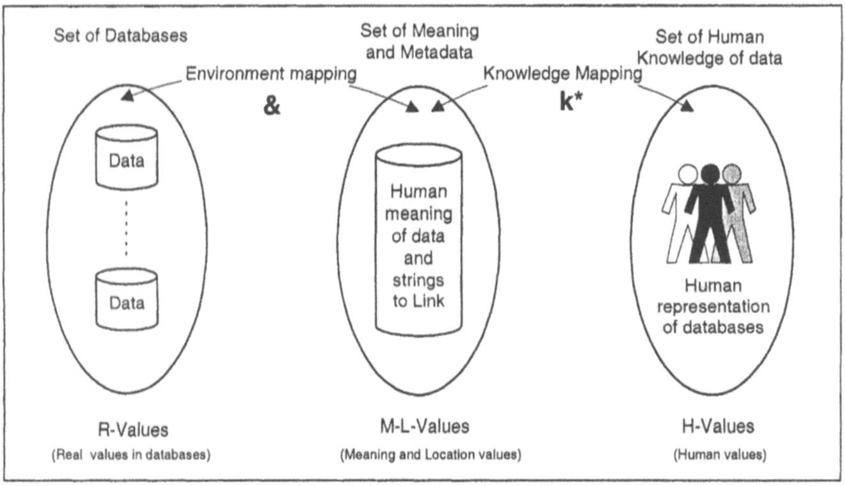

Figure 1 – Environment Model

Basically we must provide an automated production of what else expressions in human language world that match the complex terms which builds the complex and understanding industrial data environments. This information is formally represented (De Roeck et al, 1991) as a set of equivalences of the form:

$$\pi(\upsilon,\varsigma,....)\longleftrightarrow\phi \text{ Where:}$$

The left part corresponds to a particular definition that we want to give lexical meaning to the database, i.e Human-Values (*H-value*). The right part corresponds to a DRC expression, defining how to and where to retrieve meaning in the collection of tables and attributes of the specific databases, i.e. Meaning and Location values (*M-L-value*).

These expressions correspond to human words, which reflect particular worlds where the human lexicon can be used. A particular user can use a different human lexicon to represent the same query as another with different lexicon. Similarly some lexicon has a meaning in some database, but in other databases could have a different meaning. The implemented test helps to identify the lexical meaning of human words within a specific database schema. This corresponds to customise a specific industrial environment (back end database schema) to a specific world human language.

The documentation system gives a set of correct set of metadata-linked expression for a specified database associated with the definition of some human language. To do that, we use the Domain Relational Calculus (DRC). Then we use *domain* variables that take on values from an attribute *domain*, rather than values for the entire *tupple* (row). An expression in the DRC is of the form (Korth, 1991):

$\{< x_1, x_2, \ldots, x_n > | P(x_1, x_2, \ldots, x_n)\}$ where $x_1, x_2, \ldots, x_n$ represent domain variables. P represents a formula composed of atoms.

2.1. Safety of the expressions

In (Korth, 1991), an expression $\{< x_1, x_2, \ldots, x_n > | P(x_1, x_2, \ldots, x_n)\}$ is safe if all of the following rules hold:

Rule 1 All values that appear in *tuples* of the expression are values from *dom(P)*.

Rule 2 For every there "exists" sub formula of the form $\exists x(P_1(x))$, the sub formula is true if and only if there is a value *x* in *dom(P1)* such that *P1(x)* is true.

Rule 3 For every "for all" sub formula of the form $\forall x(P_1(x))$, the sub formula is true if and only if *P1(x)* is true for all values *x* from *dom(P1)*.

A higher problem remains unsolved: it is the human language (h-value) expression correctly expressed over the defined database application via the Extended Data Model (EDM) mapping? The customizing tool cannot evaluate the correctness and safety of the *h-value* expression. The tool helps to give a particular meaning related to the specific industrial environment. The documentation tool will supply by other hand, correct mapping expressions.

Formulae are: $h-value(X) = \Im(r-value)$

where $\Im$ is the function that returns the address where the *h-value* has meaning. This function is & and it will use: *Meaning* and *Location* values. Each one, corresponds:

- *Meaning (X)* = phrase that specify meaning of the contents data of a specific table/field
- *Location (X)* = Technique (ODBC driver) used to access that type of data (Sanders, 1998)

If we have this mapping, we can also have the inverse function:

$$(r-value)^{-1} = \Im^{-1}(h-value) = k^*(r-value)$$

2.2. Operations

Using a first logic order calculus and a DRC, we must supply a set of operations that fulfill our goal: mapping *h-values* to *r-values* using *m-l-values*, in the industrial environment.

2.2.1. Relating a word with some domain/table – *Select*

$$h-value(x_1,...,x_n) \Leftrightarrow < y_1,...,y_m > \in r$$

2.2.2. Relating a word by the *union* of relations

$$h-value(x_1,...,x_n) \Leftrightarrow < y_1,...,y_m > \in r_1$$

$$h-value(x_1,...,x_n) \Leftrightarrow < y_1,...,y_m > \in r_2$$

2.2.3. Other operations

Other operations must be supplied, to complete all the overall set of operations:

- Using the Existential Quantifier:
  $h-value \Leftrightarrow \exists variable(DRC_1 \wedge DRC_2 \wedge ... \wedge DRC_n)$
- Definition involving join of relations
  $h-value(X,Y) \Leftrightarrow r_1(X)\, join\, r_2(Y)$
- Definition involving the existence of some constrained constant
  $h-value(x_1,...,x_n) \Leftrightarrow \exists c_1,...,c_m(< x,c_j > \in r) \wedge c_j \Theta k$
- Constraints on the rules variables
  $h-value(x_1,...,x_n) \Leftrightarrow x_i \Theta x_j$
- Definition involving *cascaded* joins
  $h-value(x_1,...,x_n) \Leftrightarrow \exists a_1(\exists a_2(...(\exists a_m(< x_i,a_j > \in r) \wedge ... \wedge < x_k,a_i > \in r_p))...)$
- Definition involving *anding* joins
  $h-value(x_1,...,x_n) \Leftrightarrow \exists a_{11},...a_{1n}(< x_i,a_j > \in r_1) \wedge ... \wedge < \exists x_{n1},...,a_{nm}(< x_k,a_i >) \in r_p)$

3. METAQUERIES

After the documentation process of the industrial environment, users must have access to identified data, which must be access using SQL commands or other tasks. This is done using metaqueries (Shen et al, 1996)(Angiulli et al, 2000).

This system creates an environment that assists the final user in the creation of queries to different DBMS. This assistance is made at two levels: first level is the translation of the business query to the database query; second level is the simplification of information retrieval. This is a query-parsing algorithm that has the following tasks:

1) Check the query syntax
2) Parsing the query to obtain all the information about tables, fields, databases and conditions.
3) Fill out missing information in the query (tables that don't refer directly to databases, ...)
4) Transform all the elements of the business query to the correspondent elements of the real query.
5) Divide the query in sub queries to sent to each of the different DBMS
6) Collect the results of the several queries executed
7) Combine the results by all the filtering conditions between fields of the different databases
8) Transform all the field names from the real field name to the business name
9) Return business information to thc end user.

4. IMPLEMENTATION SYSTEM

The model is build around the ideas exposed previously. An implementation model is based on Larry P. English (English, 1999) and Fayyad (Fayyad et al, 1996).

Outsider systems, that use this data (human understand of data (*h-value*)) are responsible for building queries, which will be transformed, by rules, on metadata to a real query on databases.

In the implemented model, Exploration, Documentation and Migration modules are implemented, using Java (Visual Café 2.0® from Sysmantec).

4.1 Exploration Module

Exploration module builds a catalogue from the industrial environment databases systems. This catalogue is based on the importation of metadata from the original database system.

Typically, we concentrated on metadata description, using the following algorithm:

```
Configure ODBC
User select database and tables
           For each table
               If it doesn't exist on meta data database, add it in
               Store number of rows with current date
               For each field
                      Store all relevant information on metadata
                      database
Store also the parameters of ODBC driver that was used
```

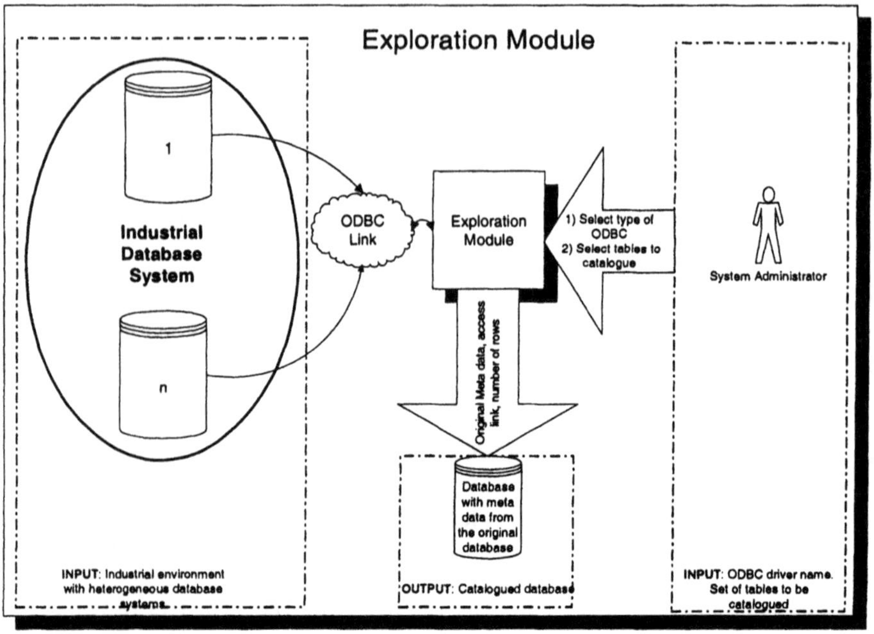

Figure 2 - Exploration Module

4.2 Documentation Module

Creates an abstraction of the metadata structure, more precisely fields and tables names. The main point is to simplify the information structure all over the existing data. The goal is to define a human value (meaning) for that data. This meaning values will be used in the next module to help user preparing the relevant information.

This module is connected to the real data, and the user creates an abstraction of the industrial environment, giving to it a human meaning. Results of this human activity are stored in a special database, for future use.

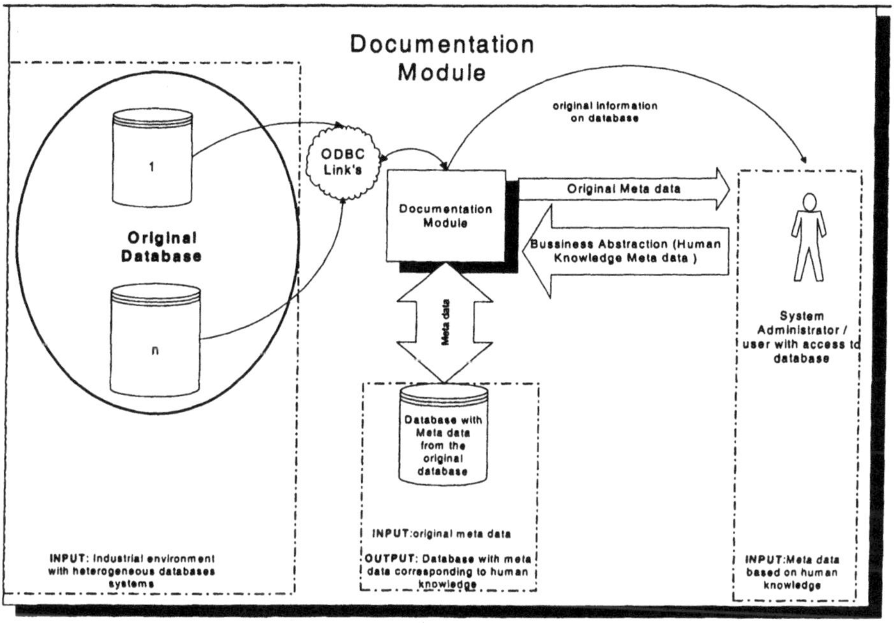

Figure 3 - Documentation Module

4.3 Migration Module

This module is the link from the selection data and the outsider users. This module must generate the final data structure, which is human values, mapped to the real data. It will use the metadata, read from the database environment.

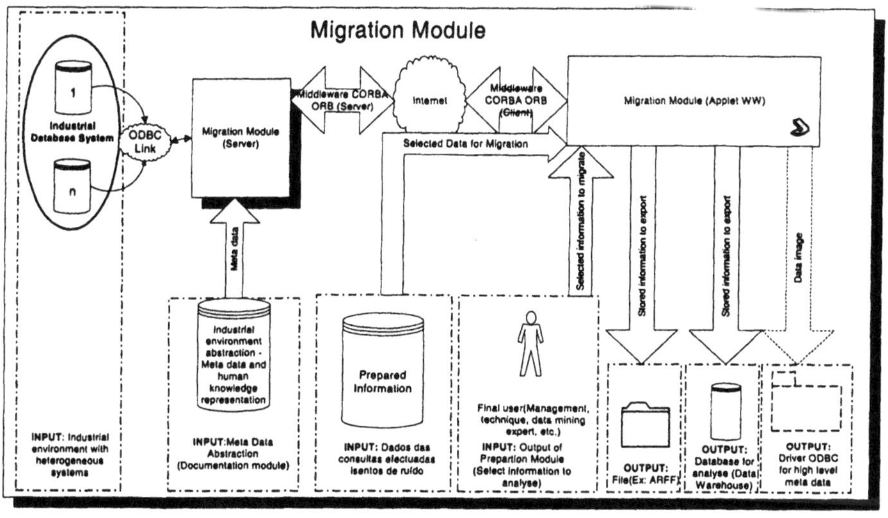

Figure 4 - Migration Module

5. RESULTS AND OUTLOOK

We have implemented and tested this model in a real production enterprise in the area of apparel for man. We implemented it with JAVA (OMG, 1996)(OAG, 1999)(OAG, 2000)(Witten et al, 1999) (Visual Café 2.0© from Sysmantec), in the following industrial environment: 1-HP9000 K250 with HP-UX 10.02© with two databases in Informix© SE 7.01, 1-HP©LC2000-biprocessor with seven databases running MS/SQL Server 7.0©. There are over 2000 tables occupying more than 1,5Gb. This structure contains also databases in two languages: German and Portuguese.

Figures 5 shows applets from the implemented tool. Figures 6 shows some obtained results.

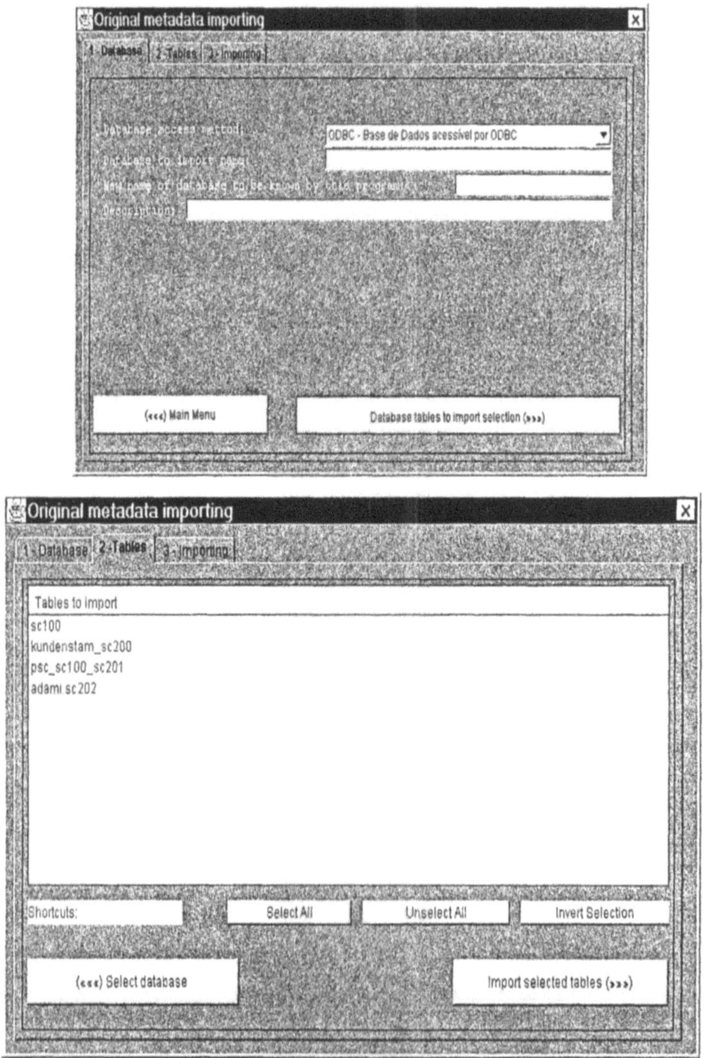

Figure 5 - **User Interface screen**

Exploration module was used to catalogue all the databases of the enterprise. We have also implemented (on request) a new module to obtain some statistical values from the databases. Some examples are the database growing rate, used to predict database allocation disk space that is useful to detect future disk upgrades, and the record count differential sampling, used to detect base and operational tables. We are able to answer to also the following questions: what is the relation between used and not used tables? We can sort tables using occupation and data access? What is the growing rate for every table?

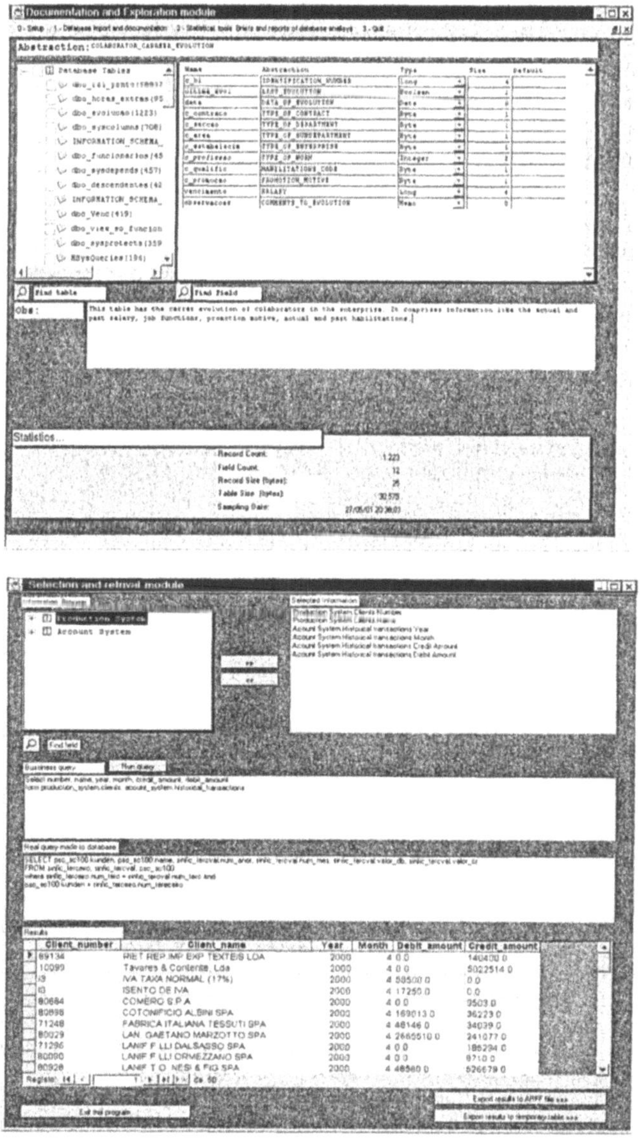

Figure 6 - Some results

6. CONCLUSIONS

The advantage of this system resides on the simplification that it provides into access and understanding industrial (or other) data environments. It is easy to use data from different languages, since we can translate meaning of data, without change the result access method.

Both implemented modules, are indispensable tools to catalogue and to document a real industrial environment. With statistical complements, we are able to help some administrator and engineering activities, which are very important to other knowledge activities.

7. FUTURE WORK

We must provide a higher security access to the data, since we can access databases over the internet (OAG, 2000)(Kahng, 2001).

Remaining modules will be implemented: preparation and cleaning as well as the development of KDD modules with interactivity (Hirji, 2001)(Fayyad, 1996).

Acknowledgments

The authors thank the support given by DIELMAR, Sociedade de Confecções S.A. for the contribution allowing the implementation the prototype system in their systems.

8. REFERENCES

1. Angiulli, F., R. Ben-Eliyahu-Zohary, L. Palopoli, G. B. Ianni, Computational properties of metaquerying problems, Proceedings of the nineteenth ACM SIGMOD-SIGACT-SIGART symposium on Principles of database systems, Pages: 237 - 244, ACM Press, 2000
2. De Roeck A.N.; Jowsey H.E.; Lowden B.G.T.; Turner R. and Walls, B.R., Natural Language Front Ends: A Formal Semantic Approach, in: Internal Report CSM/138. University of Essex, Department of Computer Science, 1991
3. English, Larry P., Improving Data Warehouse and Business Information Quality – Methods for reducing costs and increasing profits, John Wiley & Sons Inc., 1999
4. Fayyad, Usama, Gregory Piatetsky-Shapiro, Padhraic Smyth, The KDD process for extracting useful knowledge from volumes of data, Communications of the ACM, Volume 39, Issue 11, November 1996
5. Hannus, Matti, Evolution of IT in construction over the last decades, http://cic.vtt.fi/hannus/, 1998
6. Hirji, Karim K., IBM Canada Ltd., Markham, Ont., Canada, Exploring data mining implementation, Communications of the ACM, Volume 44, Issue 7, Pages: 87 - 93, July, 2001
7. Kahng, Andrew B., Design technology productivity in the DSM era, Proceedings of the conference on Asia South Pacific Design Automation Conference, Pages: 443 - 448, Series-Proceeding-Article, ACM Press, 2001
8. Korth, F. H, Silberschatz, A., Database System Concepts, McGraw Hill International Editions, Computer Science Series, p. 443-471, 2nd Ed. 1991
9. Laudon, Kennet & Jane, Essentials of Management Information Systems (Transforming Business and Management) International Edition – Third Edition - Prentice Hall International Editions, 1999
10. Object Management Group - OMG, Manufacturing Special Interest Group - SIG, Manufacturing Enterprise Systems - A white paper, OMG, Document MFG/96-01-02, Version 1.0, http://cgi.omg.org/cgi-bin/doc?mfg/96-01-02, 1996

11. Open Applications Group, Common Middleware API Specification, Document OAMAS_990815, Open Applications Group, Inc., 1999
12. Open Applications Group, Plug and Play Business Software Integration, The Compelling Value of the Open Application Group, White Paper, Open Applications Group, Inc., 2000
13. Sanders, Roger E., ODBC 3.5 Developer's Guide, McGraw-Hill Professional Publishing, 1998
14. Sargent, Philip, Product Data Management (PDM) and Manufacturing Resource Planning (MRP), PDM Articles, The Management Magazine for Mid-Sized Manufacturers, Product Data Management Information Center (PDMIC) http://www.pdmic.com/articles/pdm_mrp.shtml, 1997
15. Shen, Wei-Min, KayLiang Ong, Bharat Mitbander, Carlo Zaniolo, "Metqueries for Data Mining" in Advances in Knowledge Discovery and Data Mining, Edited by Usama M. Fayyad, Gregory Piatetsky-Shapiro, Padhraic Smyth & Ramasamy Uthurusamy, AAAI Press/The MIT Press, 1996
16. Staud, M., J.-U.Kietz, U.Reimer, "A Data Mining Support Environment and its Application on Insurance Data", in Proceedings The Fourth International Conference on Knowledge Discovery & Data Mining, 1998
17. Witten, Ian H., Eibe Frank, Data Mining – Practical Machine Learning Tools and Technologies with Java Implementations Morgan Kaufmann Publishers, 1999

A FRAMEWORK FOR UNDERSTANDING CELLULAR MANUFACTURING SYSTEMS

Sílvio do Carmo Silva, Anabela Carvalho Alves
Minho University, Campus de Gualtar, 4710 – 057 – Braga, Portugal
phone:+351+253 604745, fax: +351 253 604741, e-mail scarmo@dps.uminho.pt
Minho University, Campus de Azurém, 4800 – 058 – Guimarães, Portugal
phone:+351+253 511670, fax:+351 253 510343, e-mail anabela@dps.uminho.pt

Many practical benefits, such as superior quality of products and short manufacturing lead times, are usually associated with Cellular Manufacturing. These and other benefits can lead to important competitive advantages for companies. However, to fully achieve these benefits there is a need for an evolution from the traditional concept of CM to the more comprehensive one, which we call Product Oriented Manufacturing. Here, systems are dynamically reconfigured for the total manufacturing of complete products, not parts only.

In this paper, we make a contribution to better understanding the nature of cells and POM Systems. Thus, a classification framework is presented of the different types of cells that might be formed and considered as building blocks for POMS.

Keywords: Cellular manufacturing, operational and physical configurations

1. INTRODUCTION

Traditionally a manufacturing cell has been identified as a system dedicated to the manufacture of a family of identical parts. The manufacture based on a setting of such cells is usually referred to as Cellular Manufacturing.

A more comprehensive definition of a *manufacturing cell* points to a manufacturing system that groups and organizes the manufacturing resources, such as people, machines, tools, buffers, and handling devices, dedicated to the manufacture of a part family, or the assembly of a family of products, with identical or very similar manufacturing requirements. Therefore, important economies of scale can be obtained by producing for economies of scope, i.e. for a variety of products.

This approach of identical or very similar processing of similar objects is known as Group Technology (GT) (Gallagher, 1973). It is for this reason that manufacturing systems based on cells are frequently associated with GT.

The problems to be solved in Cellular Manufacturing Systems (CMS) can be classified as cell design and cell operation problems.

Arvindh and Irani (1994) identified four classes of problems to be solved in the design of cellular manufacturing cells, namely: *machine group and part family formation, machine duplication, intra-cell layout and inter-cell layout.* These authors argue that such problems are closely interrelated and must be solved in an integrated and iterative manner. They go on to propose a method for cell design based on this integrated approach. Nevertheless, most methods and models that have been proposed only solve one or some of the identified problems in a manner that does not take into consideration all the interrelationships referred to [(Suresh, 1998), (Moodie, 1995)]. Due to this, the results from such methods or models tend to be somewhat inefficient and, in many cases, of little practical value.

In addition to the design problems pointed out by Arvindh and Irani, operation problems must be solved. These have to do mainly with production control including scheduling. For these a variety of methods have also been identified [(Suresh, 1998), (Moodie, 1995)].

When designing CMS, the need for parts production coordination for making complete products or meeting customer orders of end items is rarely taken into consideration. Thus, the need for a quick response to customer requirements, which is recognized as an important strategic objective, is not taken explicitly into full account. This limitation, however, has been addressed in recent years through a variety of systems interlinking a number of cells. A paradigmatic example of this is the Quick Response Manufacturing System as referred to by Suri (1998).

2. GENERIC CONFIGURATIONS OF MANUFACTURING SYSTEMS

Traditional Manufacturing Systems may be classified into two major classes, see *Figure 1*: Functional Oriented Manufacturing Systems (FOMS) and Cellular Manufacturing Systems (CMS). The former are characterized by their functional organization, frequently also referred to as process organization (Burbidge, 1996). Here, work flows between functional departments or shops according to processing requirements. Cellular Manufacturing can be seen as a simplified form of product organization that carries out production in manufacturing cells.

Usually a variety of different jobs, i.e. parts and components, or even one-of-a-kind jobs, are concurrently manufactured in FOMS in a complex network of workflows. Due to this and frequent system overloading for improving system utilization, work in process and manufacturing lead times are usually high and achieving due dates can be difficult and unpredictable. Trying to reduce drawbacks of FOMS through inventories can be acceptable in some cases but this can be costly in varying and dynamic customer demand environments.

CMS are organizationally very different. For achieving its purpose a CMS is made of complementary workstations at each stage of processing, according to the need of the family of similar jobs to be manufactured.

Important similarities of jobs of a family, in the context of CMS, have to do not only with the share of same manufacturing resources and processes but also commonly with similarity of sequence of operations, materials, geometry, tolerances and volume of production. This makes the manufacturing, including handling, transport and storage, of apparently different jobs identical.

Due to the product oriented nature of CMS, not only higher productivity, lower WIP, lower throughput time and better production control can be expected but also higher volumes of production can be better dealt with than within FOMS. Because of this, CMS can be seen as building blocks for what Skinner (1974) defines as a focused factory.

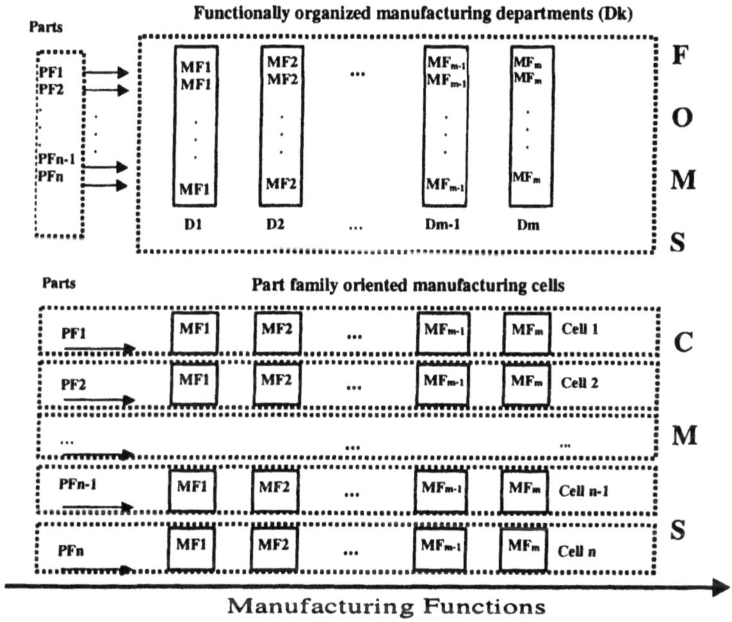

PFi - the similar set of parts to manufacture or products to assemble of family Fi

Figure 1 – A generic and interrelated view of a FOMS and a CMS

3. CONCEPTUAL MANUFACTURING CELL CONFIGURATIONS

Five different flows of work can be identified in a manufacturing cell, namely direct, direct with bypassing, inverse, inverse with bypassing and repetitive, see *Figure 2*.

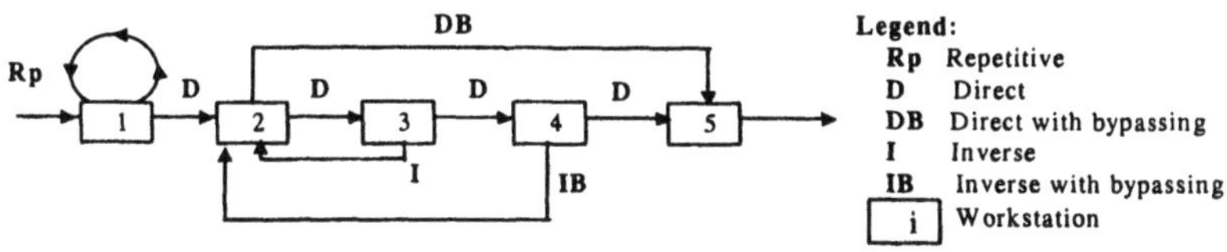

Figure 2 – Workflows in manufacturing cells (adapted from Aneke, 1986)

Repetitive flow means that the same workstation is required in sequence two or more times by the same job. It is the combination of these flows that differentiates cell configurations.

3.1 Basic Cell Configurations

Workflow in cellular manufacturing should preferably be direct, with every job having exactly the same number of operations. We name a cell with these characteristics as a Pure Flow Cell (PFC). If the number of operations is not the same for all jobs processed in direct flow and, additionally, bypassing flow is allowed, then we could say that we have a more general concept of flow cell, and call it General Flow Cell (GFC). In the more general case we can consider unrestricted direction of the flow of work and call the cell General Cell (GC). Therefore, not only direct with bypassing, but also backtracking or inverse and repetitive flow can be allowed. This GC cell case applies when both the number of operations and operations sequences are not identical for all parts of the family. Clearly this situation is more difficult to manage and exploit than the much simpler direct flow cell, with or without bypassing flow.

In addition to the aforementioned cell configurations we may identify the Single Workstation Cell (SWC) configuration. Clearly, from workflows shown in *Figure 2* the repetitive is the only one that might be available in such a cell. This does not exclude the possibility of workstation intra-flow, which may be seen as a repetitive flow within the same workstation, taking place between its parallel processors, if they exist.

We call the four identified configurations Basic Cell Configurations, see *Figure 3*.

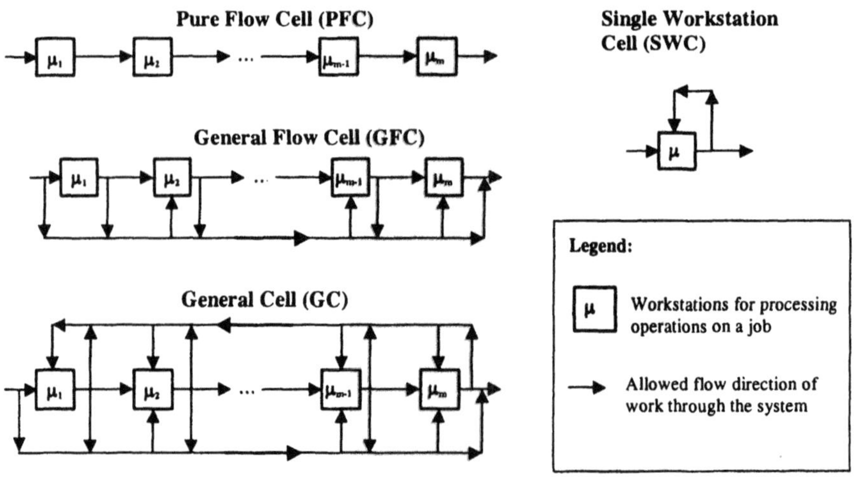

Figure 3 - Schematic representation of the four Basic Cell Configurations

These have their counterparts in the theory of scheduling. Thus, the PFC, the GFC and the GC may be respectively associated with pure flow lines, general flow lines and general shops [(Baker, 1974), (Brucker, 1995), (Blazewicz, 1996)]. Although a cell may work as a general shop, or as its job-shop instance, the tendency is to avoid such situations because this would require complex control at various levels of production.

Some authors focus only on the flow type configurations when talking about cells. They refer to cells and lines as being identical concepts. In fact the distinction between the two can be difficult to make. Such a distinction seems to be based more on matters of system operation and manning than on the nature of the flow of materials. Thus, for instance, such difficulty is underlined by Schonberger (1983) when he says that "... *cells and production lines are cut out of the same cloth*" and by Black and Schroer (1988) who state, comparing cells with lines, that *"the unique feature of a cell is that the cycle time is not dependent on the manufacturing time of a particular machine*". It appears that cells are different from balanced manufacturing or assembly lines. However there is no reason why we cannot have balanced flow lines manufacturing a family of parts or assembling a group of similar products, which is the same as to say balanced cells.

The SWC configuration is a more general case than single processor or parallel processor systems referred to in scheduling theory. However, a strong approximation exists. This is because the workstation may in fact be provided with a single processor or with a few parallel processors. However, as aforementioned, they behave as a more general system because repetitive work is considered and inter processor flow of materials is allowed, for instance, when more than the operation of one job is to be carried out in the same workstation having parallel processors.

3.2 Non - Basic Cell Configurations

A typical problem in the formation of cells is the insufficiency of replicated manufacturing resources, mainly machines, to be allocated to different cells. To solve the problem two strategies are usually followed. The first looks for enlarged part families as a way of sharing scarce resources, which are then allocated to fewer cells. The main aim of this strategy is to keep the manufacturing of the entire spectrum of parts of each family in its own cell, creating, in this way, independent and autonomous cells. We could say that there is a gain in autonomy but a probable loss in cell efficiency. In fact, the similarity within such enlarged families is reduced as a result, making manufacturing more complex.

The second strategy includes two logics. In one, cells share their resources with other cells, allocating load to process parts belonging to families initially allocated to other cells. In this case inter-cellular flow of parts takes place. In the other logic some resources are independent and shared by more than one cell. Irani, Cohen and Cavalier (1992) call this arrangement *hybrid layout,* considering that shared resources are organized by function. In both logics, however, we can observe that shared equipment between cells exists. So, we also interpret these cells as shared cells. The result in both cases is disturbed flow of work in any shared cell. So, the disturbing behaviour of work within cells is similar in both the aforementioned logics. This is probably the most important factor affecting operational efficiency of shared cells. Due to this, we do not distinguish between hybrid layout and shared cells, calling any cell involved in intercellular workflow Shared Cell, see Figure 4.

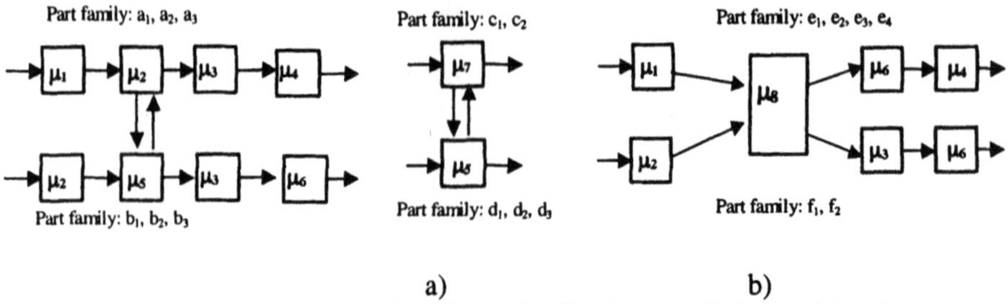

Figure 4 – Schematic representation of shared cells: a) inter-cellular workflow between shared cells b) shared cells in a hybrid layout

The set of Basic and Non-Basic cells we call Conceptual Cells, see Figure 5.

An alternative name for the Conceptual Cells could be based on the independence of processing. In this case we could say that basic configurations correspond to independent cells and the non-basic to dependent cells.

BASIC	Single Workstation Cell	SWC
	Pure Flow Cell	PFC
	General Flow Cell	GFC
	General Cell	GC
NON-BASIC	Shared Single Workstation Cell	SSWC
	Shared Pure Flow Cell	SPFC
	Shared General Flow Cell	SGFC
	Shared General Cell	SGC

Figure 5 – Conceptual Cells in Cellular Manufacturing

3.3 Virtual Manufacturing Cells

A Virtual Manufacturing Cell (VMC) is a concept which was initially reported in 1982 by Simpson, Hocken and Albus, according to Ratchev (2001), and by McLean, Bloom and Hopp (McLean, 1987). VMC can be seen as dynamic software arrangements of machines or workstations, of large facilities, to be operated under cellular manufacturing for a given time period, without re-arranging their physical layout. The advantage of virtual cells stems from the fact that dynamic virtual cell reconfiguration can easily and frequently be carried out at minimum cost. VMC are important to match changing market requirements when and where frequent alterations of the physical manufacturing system layout are uneconomical and impractical or cannot take place.

VMC may be configured as virtual cell instances of Conceptual Cells.

Ratchev (2001) put forward an approach, quite different from conventional ones, for the formation of virtual manufacturing cells, of the shared class, in a machining environment. The approach based on machine processing capabilities can define a set of virtual machining cells and the matching part families, based on a processing plan for each part. These plans, expressed in the so-called Resource Elements (RE) of the machine shop, permit alternative use of machines since more than one machine is likely to offer the same RE.

3.4 Resource Combination and Flexibility of Workstations

The identified Conceptual Configurations embrace quite a few instances that have to do with resource combination and flexibility of workstations. Thus, the nature and quantity of manufacturing resources, including main resources, such as machines, and auxiliary resources, such as operators and tools, involved in each workstation, brings about different instance types of each Conceptual Configuration and poses different problems to be solved at both the design and operation levels of CMS.

Iin fact, the number of operators and the level of replicated auxiliary equipment, such as tools, together with their dynamic utilization within cells, may substantially affect not only the cell capacity and manufacturing flexibility, but also the manner that cells are operated. Therefore, auxiliary resources largely determine the performance level of manufacturing cells (Silva, 1997, 1995, 1988).

Moreover, cell efficiency and effectiveness, being highly dependent on cell operation, is also influenced by the configurations of each workstation in a cell. The configuration of workstations may take different forms according to manufacturing requirements and objectives. They may be simple, provided with a single machine to carry out a single manufacturing function or be more complex involving three other situations, which may be combined, namely having a) parallel processors, b) multiple resources or processors and c) multifunction processors.

Flow shops having *parallel processors* have been referred to as *Flexible Flow Shops* (Pinedo, 1995) or *Hybrid Flow Shops* (Elmaghraby, 1997). So we could consider a *flexible flow cell* to be a flow cell with parallel processors. Thc need for parallel processors may be justified mainly on capacity, operational reliability and flexibility grounds.

Often work must be carried out simultaneously using a number of resources at a workstation. These may include, for instance, machines, tools and operators. These systems are referred to in the literature as multiprocessor task systems for which a class of scheduling problems is identified [(Brucker, 1995), (Blazewicz, 1996)]. Again, instance types of Conceptual Cell Configurations having multiprocessor tasks can be considered which may be called *multiprocessor task cells.*

Furthermore, workstations may be provided with flexible machines capable of carrying out a range of different operations, usually through changing manufacturing aids such as tools, numerical control programs and fixtures. This is typical in Flexible Manufacturing Systems and Cells (Silva, 1988). Cells with flexible machines can be named *multifunction processor cells.* A class of scheduling problems for Multifunction Processor Systems is identified in the literature (Brucker, 1995).

From the above we can conclude that taking into consideration the characteristics of workstations we may arrive at several variations on the Basic and Non-basic Conceptual Cell Configurations. Moreover, if flexibility is explored, through dynamic utilization of manufacturing resources, and autonomous cells are interlinked to achieve coordinated production of end items, we may arrive at several system concepts widely referred to and used in practice, which we classify under the heading of Product Oriented Manufacturing Systems (POMS).

4. PRODUCT ORIENTED MANUFACTURING SYSTEMS - POMS

Although CMS can have a beneficial impact on manufacturing operations of an enterprise, the full benefits of such a product-oriented approach to production can only be realized when overall production is considered. This means that good production of parts or the assembly of products alone does not necessarily mean effective advantages for a company as a whole. It is important that customers' full orders are quickly satisfied with high quality and good use of manufacturing resources. To effectively answer these challenges, CMS must evolve into POMS. A POMS system is a set of interlinked manufacturing resources or cells that simultaneously addresses the manufacturing of a certain ordered quantity of a product, or of a range of similar products, in a coordinated manner. A product may be simple, like a part, or complex, having a product manufacturing and assembly structure, involving several levels. This may be related with the multilevel bill of materials (BOM) of the product. When the product is simple, a POMS may simply take a form of a cell. Otherwise, it may take the form of a coordinated set of interlinked cells.

This coordination of work among manufacturing resources or cells is one of the most distinguishing aspects of POMS. A set of cells, which does not work under such a coordination setting across the requirement of all items and tasks of the overall and final product, is not a POMS.

An enlarged view of the POMS concept includes logistic operations, mainly when production resources are distributed in space. To ensure maximum success, production under this concept must be able to fully and dynamically consider and involve resources available to a company, over a time period, locally or globally, either belonging to its own or to potential production partners.

Dynamic reconfiguration of POMS under changing market requirements is most probably necessary. This necessity is also justified because of the dedicated nature of POMS to a specific mix of products which, changing over time, calls for new arrangements to ensure high levels of operational performance.

POMS may be built by putting together, in a localized site, manufacturing means or resources that may be physically dispersed or alternatively, when such is not possible or is uneconomical, by organizing them in virtual POMS. For this a methodology should be applied that deals, in an integrated and iterative manner, with the many problems of POMS design. Silva and Alves (2001) have made a contribution in this direction.

Quite a few variations of the POMS concept have been put forward in recent years. All of them may derive their configurations from the Conceptual Cells presented in this work. Some of the proposed and actual POMS draw upon a range of technologies and methods not available in the past, in domains such as computer science, information systems, electronic communications including intranets and Internet, mechatronics and company internal and external logistics. Such systems, some of which are very familiar in practice, are known under names such as focused factory, just-in-time, lean, linked-cell, OPIM (One-Product-Integrated-Manufacturing), quick response, flexible, reconfigurable and virtual systems [(Silva, 2001), (Monden, 1983), (Lewis, 2000), (Black, 1991), (Suri, 1998), (Putnik, 1995), (Chen, 1998) (Tempelmeier, 1993), (Mehrabi, 2000), (Drolet, 1996), (Lee, 1997) (Iwata, 1995), (Camarinha-Matos, 1999) and (Ratchev, 2001)].

5. CONCLUDING REMARKS

To keep up with the increasing and changing market demands of today and tomorrow, companies must be able to efficiently manufacture and quickly deliver good quality products to customers. To achieve this today companies cannot rely on traditional organization and operation of systems based on functional departments. Moreover, cellular manufacturing based on uncoordinated or loosely coordinated manufacturing cells is also inappropriate. Present requirements indicate that a more holistic approach to manufacturing is necessary. This can be achieved through Product Oriented Manufacturing Systems (POMS) dynamically built of cells or manufacturing resources, local or globally available, which must be interlinked and closely coordinated for the total production of complete products, not parts only. These products may preferably bear manufacturing similarities. Although a localized physical set-up for such a purpose should be sought, POMS are likely to be more dependent on virtual reconfiguration when resources are dispersed or are uneconomical to rearrange.

Faced with apparently different recent and past manufacturing systems, such as focused factory, just-in-time, lean, linked-cell, OPIM, quick response, flexible, reconfigurable and virtual systems, we can observe that, in different degrees, they all have the identical characteristic of being product oriented systems as defined in this work. This is why we consider them to be instances of the general POMS concept.

A framework of conceptual cells is presented in this work, which can contribute to a better understanding of cellular manufacturing as instrumental for helping in the design and operation of POMS. This framework reduces cells to two sets, the basic and non-basic cells, for which virtual counterparts can be considered. Furthermore, the nature of cell workstations is explained by taking into consideration manufacturing resource combination and flexibility.

6. REFERENCES

1. Aneke NAG, Carrie AS. A design technique for the layout of multi-product flowlines. International Journal of Production Research, 1986; 24:471-481.
2. Arvindh B, Irani SA. Cell formation: the need for an integrated solution of the problems. Int. Journal of Production Research, 1994; 32: 5.
3. Baker, KB. Introduction to Sequencing and Scheduling. John Wiley and Sons, 1974.
4. Black, JT. The Design of the Factory with a Future. McGraw-Hill, 1991.
5. Black JT, Schroer BJ. Decouplers in Integrated Cellular Manufacturing Systems, Journal of Engineering for Industry, 1988; 110:77.
6. Blazewicz J, Ecker KH, Pesch E, Schmidt G, Weglarz J. Scheduling Computer and Manufacturing Processes. Springer Verlag, Heidelberg, 1996.
7. Brucker, P. Scheduling Algorithms. Springer, 1995.

8. Burbidge, JL. Production Flow Analysis for planning Group Technology. 1ª publicação 1989, Oxford: Clarendon Press, 1996.
9. Camarinha-Matos LM, Afsarmanesh H. "The Virtual Enterprise Concept". In Working Conference on Infraestructures for Virtual Enterprises (PRO-VE'99), L. M. Camarinha-Matos and H. Afsarmanesh, ed., Kluwer Academic Publishers, 1999.
10. Chen, F. Frank. Flexible production systems for the apparel and metal working industries: a contrast study on technologies and contributions. Int. J. of Clothing Science and Technology, 1998; 10:1.
11. Drolet JR, Montreuil B, Moodie CL. Empirical Investigation of Virtual Cellular Manufacturing System. Symposium of Industrial Engineering - SIE'96, 1996.
12. Elmaghraby SE, Karnoub RE. "Production Control in Hybrid Flowshops: An example from textile Manufacturing". In The Planning and Scheduling of Production Systems, A. Artiba and S. E. Elmaghraby, ed., Chapman and Hall, 1997.
13. Gallagher CC, Knight WA. Group Technology. Butterworths, 1973.
14. Irani SA, Cohen PH, Cavalier TM. Design of Cellular Manufacturing Systems. Transactions of the ASME, 1992; 114.
15. Iwata K, Onosato M, Teramoto K, Osaki S. A Modelling and Simulation Architecture for Virtual Manufacturing Systems. Annals of the CIRP, 1995; 44:1.
16. Lee, KI. Virtual Manufacturing System - A Test Bed of Engineering Activities. Annals of the CIRP, 1997; 46:1.
17. Lewis, MA. Lean production and sustainable competitive advantage. Int. J. of Operations & Production Management, 2000; 20; 8, 959-978.
18. McLean CR, Brown, PF. "The Automated Manufacturing Research Facility at the National Bureau of Standards". In New Technologies for Production Management systems, H. Yoshikawa e J. L. Burbidge, ed., Elsevier Science Publishers B. V. North – Holland, 1987.
19. Mehrabi MG, Ulsoy AG, Koren Y. Reconfigurable Manufacturing Systems: Key to Manufacturing. Journal of Intelligent Manufacturing, 2000; 11: 403-419.
20. Monden, Y. Toyota Production System. Industrial Engineering and Management Press, Institute of Industrial Engineers, 1983.
21. Moodie C, Uzsoy R, Yih Y. Manufacturing Cells – A System Engineering View. Taylor & Francis, 1995.
22. Pinedo, M. Scheduling – Theory, Algorithms and Systems. New Jersey: Prentice-Hall Inc, 1995.
23. Putnik GD, Silva SC. One-Product-Integrated-Manufacturing. In Balanced Automation Systems, L. M. Camarinha-Matos, H. Afsarmanesh, eds. Chapman & Hall, 1995.
24. Ratchev, SM. Concurrent process and facility prototyping for formation of virtual manufacturing cells. Integrated Manufacturing Systems, 2001; 12: 4, 306-315.
25. Schonberger, RJ. Plant Layout Becomes Product-Oriented with Cellular, Just-In-Time Production Concepts. Industrial Management, 1983, 66-70.
26. Silva, SC. An Investigation into tooling requirements and strategies for FMS operation. PhD Thesis. LUT, UK, 1988.
27. Silva SC., Alves, AC. SPOP - Sistemas de Produção Orientados ao Produto. Células Autónomas de Produção - TeamWork'2001 Conferencia, Lisboa, 2001.
28. Silva SC, Putnik GD. "Gestão de Produção - Influência dos Meios Auxiliares na Flexibilidade Operatória dos Sistemas Flexíveis de Manufactura", Seminário COMET - Gestão da Produção em Tempo Real na Indústria do Vestuário, Universidade do Minho, Guimarães, 1995.
29. Silva, SC. "Analytical Assessment of Tooling Requirements for FMS Design and Operation". In Reengineering for Sustainable Industrial Production, L. M. Camarinha-Matos, ed., Chapman & Hall, 1997.
30. Silva SC, Alves AC. Uma Metodologia para o Projecto de Sistemas de Produção Orientados ao Produto. VII Int. Conf. on Industrial Engineering and Operations Management, Brasil, 2001.
31. Suresh NC, Kay JM. Group Technology and Cellular Manufacturing – State of the Art Synthesis of Research and Practice. Kluwer Academic Publishers, 1998.
32. Skinner, W. The focused factory. Harvard Business Review, 1974.
33. Sury, R. Quick Response Manufacturing – A Companywide Approach to Reducing Lead Times. Oregon: Productivity Press, 1998.
34. Tempelmeier, H, Kuhn, H. Flexible Manufacturing Systems. John Wiley & Sons, Inc., 1993.

SALES PREDICTION WITH MARKETING DATA INTEGRATION FOR SUPPLY CHAINS

Daniel Gillblad
Anders Holst
Swedish Institute of Computer Science
Box 1263, SE-164 29 Kista, Sweden
dgi@sics.se, aho@sics.se

Predicting customer demand, manifested as future sales, is an important task for supply chain optimisation. The ability to predict future sales can be used to do more efficient resource allocation and planning, avoiding large stocks and long delays in delivery. Sales prediction is inherently difficult, and can benefit greatly from including other information than historical sales data. Here the forecasting module of the DAMASCOS project, D3S2, will be presented, which both incorporates marketing information and customer demand in the sales prediction as well as supplying the user with options and information that makes the prediction much easier to use in practice.

Sales prediction, Supply Chain Management, Marketing Data

1. INTRODUCTION

An important component of a supply chain management system is the ability to predict the flow of products through the network and to use this information to avoid large stocks or delays in delivery. While such a tool can potentially be constructed in many ways, the central issue is to predict the future sales of the products. Sales prediction is usually a difficult task, where there might be severe problems finding relevant historical sales data to base a prediction on. Therefore, the use of all relevant information other than the historical sales records becomes important. There are several factors that may affect the sales. Most notable are sales activities, advertisements, campaigns and brand awareness of the customers, but also weather, trends, and the emergence of competing products may have a substantial impact. Some of these are very difficult to encode for the use in a prediction system, but others, like brand awareness, marketing investments and sales activities have a clear and direct impact on the future sales while being possible to represent in a reasonable way. Here, some aspects of the forecasting module of the DAMASCOS suite is presented, which tries to incorporate this information in the predictions while giving the user information that makes practical usage of the prediction easier. Some important general considerations of sales prediction for supply chain management are also discussed, and some methods for sales prediction are presented briefly.

DAMASCOS is an R&D project supported by the EC, and aims at supporting optimisation and improved information flow in the supply chain. The DAMASCOS software contains modules for sales management, distribution and logistics, supply chain coordination, integration and forecasting. The forecasting module of the DAMASCOS suite, D3S2 (Demand Driven Decision Support System), is a configurable forecasting engine that is integrated with the rest of the DAMASCOS software.

2. SALES PREDICTION FOR SUPPLY CHAIN MANAGEMENT

2.1 Sales prediction for decision support

Sales predictions will most likely be used by sales managers, sales agents, marketing managers and the persons responsible for production planning. A sales manager can use the predictions to get an indication of whether or not the sales target is going to be achieved in order to increase or redirect the sales efforts if needed. The sales agents can use the predictions to e.g. promise better delivery times on some products. A marketing manager will be able to use the predictions to early see the effects of marketing strategies and to get an early warning if a change in strategy or additional marketing investments are needed to meet the goals set by the company. The sales prediction is also very useful for the persons responsible for production planning, since capacity can be reserved at the suppliers at an early stage, raw materials can be ordered earlier than otherwise and production scheduled longer in advance.

The sales predictions from the D3S2 module are intended to be used to support manual decisions. In general, sales predictions should not be used in automatic decision processes because of the inherent uncertainty in any prediction method. Only a human operator can decide whether a prediction is plausible or not since it is impossible to encode and use all background information within a prediction system. This means that the design of the module was targeted towards creating a system that delivers as rich and accurate information as possible to its users, both regarding the actual predictions as well as what they are based on. Talking to the future users of the system, this is much appreciated.

In discussions and market surveys, many people express scepticism about sales prediction, comparing it to stock market prediction and questioning whether it is at all possible. Although this is a sound reaction, it builds on the misconceptions that sales prediction generally faces the same problems as prediction of the highly volatile stock market, and that the prediction should provide a highly accurate number of future sold goods instead of being a useful tool that can give an indication of what to expect. This means that a sales prediction module must try to give the user a feel for when to trust the prediction as well as tools that enables full use of this prediction.

Looking at a prediction alone, without any further information, it is very difficult for the user to determine how reliable that prediction is. If an important decision is to be taken partially based on that sales prediction it is necessary to know the accuracy. Therefore, the D3S2 module provides not only the predicted value but also an uncertainty measure to give the user an indication of the expected prediction error.

To further make it easier for the user, the module will provide a short explanation of what the sales prediction is based on and how it has been calculated. If the historical data used to make the prediction differs greatly from the situation at hand, the user can choose to disregard a prediction that looks good otherwise. The explanation could also help the user decide whether or not to trust the prediction when something out of the ordinary has occurred. Both this explanation and the uncertainty measure are included to increase the users trust in the prediction, and to give a more transparent feel to the module. The user should not be left thinking of the prediction as a black box whose inner workings are concealed, since this greatly reduces the usefulness of the prediction. Also, whenever possible a plot showing the development of the prediction in time as more data becomes available will be presented to the user so that the relevance of the prediction can be more easily evaluated.

2.2 Important considerations

Historical sales data is usually quite noisy and may not be very representative for a company's current and future sales. A prediction model should take this into account, regarding recent sales and marketing events more relevant than older ones, while using as much domain knowledge as possible. The number of parameters in the model must also be kept to a minimum, giving a less accurate prediction on the historical data used for estimating the model but providing for much better generalization and predictions of future sales.

In addition, in many cases the rapid changes of a companies range of products gives a need for a human operator to specify relationships to earlier products to make it possible for the prediction to use historical data. When a new article is introduced, the prediction module might benefit from knowledge about what older articles it resembles, i.e. what older articles to extract data from and use as the basis for the prediction. The D3S2 module deals with this problem by grouping all articles into *categories* and *segments*. This grouping is made by the user when entering a new item into the database. Categories and segments can be overlapping, and a prediction can be made on all articles, a category, a segment or a combination thereof. No predictions are made on a single article. The user can of course reserve a special category or segment for one article only, but must be aware of the statistical implications of doing so. If the article have been produced and sold for a long time with a decent volume, though, the prediction will probably have enough historical data to do a reasonable prediction.

2.3 Integrating the customer demand in the prediction

Integrating customer demand in the sales prediction can greatly improve the prediction results. Reliable estimation of customer demand is not very easy though. A simple approximation is to use the media investments made by the company, which can be viewed as an indicator of future customer demand since all marketing investments generally increase the awareness of the brand. Media investments can be acquired directly from the marketing department of the company. To find the impact of these marketing activities, a method called *brand tracking* can be used. The tracking continuously measures for example advertising awareness and considered brand amongst customers using telephone interviews with a statistically representative set of people, who answers questions about what brands they have noticed advertising for lately and which trademarks they prefer. If tracking data is available, it is a more reliable indicator of customer demand. It should then be used instead of media investments in the prediction, but both tracking data and media investments can be integrated in the same way in the prediction. Unfortunately, tracking data is not available in many SMEs.

It is very difficult to separate advertising or brand awareness effects on certain articles or groups of articles. The customer demand data must therefore be viewed as having a general impact of the future sales of a brand. In the D3S2 module, customer demand data is used together with historical sales data to estimate the general trend of all sales of a company or brand. This general trend is then used as a basis for all predictions by the module.

3. THE D3S2 FORECASTING MODULE

3.1 D3S2 in the DAMASCOS suite

The D3S2 module is a configurable forecast engine that runs on a centrally located server, most likely at the principal producers headquarter. This forecast engine responds to all prediction requests from all modules within the DAMASCOS suite, making the prediction functionality available to all users of the system without the need to locally store and update the large amount of data used to calculate the prediction. Since prediction is used in several other modules in the DAMASCOS suite, modules often used by different kinds of users, all interaction with the D3S2 module is made through those other modules so that the prediction system is well integrated into the users normal work tool.

Access to constantly updated data is very important to the prediction module, especially when doing short-term predictions. In the DAMASCOS suite, all orders are entered into the SALSA module. All new sales and marketing data that become available are provided without delay by the SALSA module to D3S2. This of course includes new orders and marketing investments, but also information about planned sales activities, start and end dates for sales campaigns and brand tracking results. New orders and sales campaign information are automatically sent to D3S2 as soon they are entered into the SALSA module. Marketing investments and brand tracking results on the other hand are usually kept in a separate system at the company, but the SALSA module will provide interfaces for entering this data as well, not making use of it itself but directly passing it on to the D3S2 module. This design keeps system maintenance costs low, since all users essentially only interact with the SALSA module and there is no need to keep the database of the prediction module up to date separately.

3.2 Sales predictions and the general trend

Very generalized, two distinct scenarios exist for how customer orders are collected. One is the *sales campaign scenario*, in which sales agents visit a predefined number of potential customers during a sales campaign with a defined start- and end date. The other is the *continuous order scenario*, in which orders arrive continuously at varying rate, either through agents visiting customers to collect orders or from customers that approach the company directly. Many companies operate with a mix of these scenarios. The D3S2 module handles both scenarios, using suitable prediction methods for each case.

Due to the need to use domain knowledge in the sales prediction, the prediction models can be expected to be fairly specialised for each type of company. Below two examples of simple but effective methods for sales prediction are explained briefly. The methods were mainly developed for the pilots of the DAMASCOS project; ATECA, an Italian clothes manufacturer that works mostly with season based sales and KYAIA, a Portuguese shoe manufacturer who only uses the continuous order scenario. Although developed for these companies, the methods can be expected to be applicable to all companies operating in a similar way.

Predicting the general trend forms the basis for both season based and continuous order prediction in the D3S2 module. Loosely speaking, the general trend is modelled using linear methods incorporating the sales dependency on time and the media investments or brand awareness at that time. Media investments and brand awareness usually have a time-aggregated effect on the sales, i.e. media investments in several weeks usually have larger impact than an investment in just one of those weeks. Therefore, the media investments and brand awareness at a certain point in time are represented as the sum over the last previous weeks with an exponentially decaying weight, lowering the significance of the media investment or brand awareness with time.

3.3 Season based sales prediction

To predict how many items that will be sold at the end of the sales campaign when the campaign has just started is obviously very difficult. Many or most of the articles sold during the current campaign might not have been available in the last sales campaign, and most articles that could be bought during the last campaign might be discontinued. Strictly speaking, there is a good chance that we have no relevant statistical information to base our prediction on if we cannot in some way make the approximation that the sales of the current seasons articles will behave like previous seasons articles. A reasonable prediction of the sales, perhaps even the best one, would actually be the last seasons total sales, with compensation for the long term trend. To make a prediction of the sales near the end of a campaign is of course much simpler, since most of the orders will already be in by that time.

In some cases, though, we have information that will make the predictions much easier. Since there often is a well-defined sales plan, we know how many customers the sales agents will visit. That means, in practice, that we actually know with a rather high certainty how many orders we will get in total during the season, since the mean number of orders from each client is very stable over the years. This information will help us to construct a simple but effective model to handle predictions in the season-based case. Note that the assumption that we fairly well know the total number of orders beforehand is not valid in all season based sales scenarios. It requires that the company operate like the pilot ATECA, with a pre-defined sales plan that is not adjusted too much during the campaign and a customer base that does not show a very erratic behavior when it comes to the numbers of orders placed. It is likely though that many companies, at least in the clothing business, fit these basic assumptions.

The prediction model itself is very simple. At a certain point of time t in a campaign, we know the number of orders so far, N_t, and the number of items sold so far, S_t. The prediction should be a sum of the orders we have got so far and the expected sales for the remaining period. Since we assume that we know the total number of orders, we know how many orders that will come in during the rest of the sales campaign, $N_r = N_{tot} - N_t$. This can be used to easily predict the future sales. From the earlier orders, we can estimate a probability distribution over the size of each order, P. Knowing the number of remaining orders, we can estimate the probability distribution of the total remaining sales by simply adding N_r distributions over the size of each order, P, together. Now when we have an estimate of the distribution of the remaining orders, we can calculate the expectation and standard deviation of that distribution. These are our sales prediction and uncertainty measure, respectively.

So how do we find the order size distribution P? The actual order size distribution can usually be expected to describe a distribution from the exponential family, most likely the gamma distribution. A Gaussian distribution would also be a reasonable fit, since it is approximately symmetric. In fact, the choice between a gamma and a Gaussian distribution is not critical. The attributes of the resulting distribution we are interested in are the expectation and the variance, and the same expressions for summing the distributions are valid both for the gamma and the Gaussian distribution.

This leads us to use a Gaussian distribution for the order sizes, since it is very easy to estimate. Here we will use simple Bayesian techniques [Cox, 1946; Jaynes, 1986, 1996] to make an estimation of the parameters in the distribution, i.e. the expectation and the variance. To start with, we assume that we more or less know nothing about the initial distributions of the parameters. We can assume a uniform prior distribution. This makes both the expectation and the variance inverse gamma distributed, and integration over the distributions leads to very simple expressions of the mean and variance. The assumption that we know nothing at all about the values of the mean and variance of the order size distribution is of course the most careful one; not assuming the data to have any special characteristics at all. But the fact is, that most of the time we have historical data available that gives us valuable prior knowledge about the distribution. We can use a simple prior information reasoning to find expressions for the expected mean and variance that includes prior information. Let us calculate a prior mean and variance, μ_p and σ^2_p, simply by using the mean and variance of the last season adjusted with the general trend. Then estimate the mean and variance in the data we have so far in the current season, μ_e and σ_e^2, using the expressions described above. Then we can write the new mean and variance estimates as

$$\mu = \frac{N\mu_e + k\mu_p}{N + k} \qquad \sigma^2 = \frac{N\sigma_e^2 + k\sigma_p^2}{N + k}$$

where k denotes a constant that is the weight of the prior. If $k = 1$, then the prior is weighted as one of the data observations, if $k = 2$ it counts as two observations and so on. Using these estimates the prediction becomes less sensitive to uncommonly large or small orders early in the season. At the end of the season on the other hand, the number of orders N are usually much larger than k, so the prior will have little or no effect on the final prediction.

Now when we have reasonable estimates of μ and σ^2, we can simply write the total sales prediction S_p and expected error E_p at time t as $S_p = S_t + N_r\mu$ and $E_p = \sqrt{N_r\sigma^2}$ where E_p is the expected standard deviation. This prediction model might be simple, but it deals with the fact that we do not have statistical data for all articles sold during the season and is relatively robust to all the noise that is present in the sales data.

The prediction model described above has two major flaws. First, it does not provide a reasonable expression of the variance in many situations. Second, it does not explicitly consider time passed since the start of the campaign. This leads to strange effects in the prediction if the sales goals are not exactly met, and the prediction keeps insisting on a higher or lower prediction, with low variance, until the absolute last day of the sales campaign. To counter this problem, we can extend the model somewhat. We are not going to do a complete derivation of the expressions here since that is quite extensive, but rather just notice that we have to consider uncertainties in more variables than before, e.g. that the remaining number of orders should also be considered stochastic. If we also assume that instead of having a known number of orders in the campaign, we have N number of maximum orders during a season with a probability q that such an order is actually placed. Then the expectation of the amount of remaining orders becomes

$$E_X = E_{n_k}\mu = \int_{n_k} n_k P(n_k \mid k)dn_k \mu = \mu(N-k)\left(\frac{q(1-p)}{1-pq}\right)$$

where n_k is the number of remaining orders of the season, k the number of orders so far, and p the fraction of the season that has passed. To get a prediction for the total sales of the season, we just add the actual sales so far to the expression above. Similarly, the final expression for the variance becomes

$$V_X = \frac{\sigma^2}{c-3}\left(c(N-k)\left(\frac{q(1-p)}{1-pq}\right) + (N-k)\left(\frac{q(1-p)}{1-pq}\right)\left(\left(\frac{1-q}{1-pq}\right) + (N-k)\left(\frac{q(1-pq)}{1-pq}\right)\right)\right)$$
$$+ \mu^2(N-k)\left(\frac{q(1-p)}{1-pq}\right)\left(\frac{1-q}{1-pq}\right)$$

This model is more complicated; at least the derivation of it, but it counters most of the problems encountered with the simplistic model and predictions are still easily and very quickly calculated. This is the model actually used in the D3S2 module for season based sales prediction.

3.4 Continuous order prediction

In the continuous case there is no fixed target for the prediction as in the season-based case. There is no sales campaign that we can predict the total number of orders for. Instead, the user must select a period of interest for which he or she wants a prediction of the sales. To be able to deal with arbitrary prediction periods the model must be very flexible, but the task is much simplified if we can build a reasonable model of the process that results in sales orders at the company. To avoid over fitting problems, this model should have a rather low number of free parameters.

Using the assumption that the sales follow a long-term trend with seasonal variations, which is valid for a wide range of companies, we can construct a simple model for the behavior of this kind of sales data. The first step is to automatically find the base frequency and phase of the seasonal variations, parameters that are already known in the season based case. Then we must find and model the shape and dynamics of this seasonal variation, using as few parameters as possible to make the model more stable.

In the simplest version, the complete model consists of a base level of sales, the general trend, with an added sinusoidal component representing the basic seasonal variation. There is a very straightforward method of finding the characteristics of the sine wave. By using the *discrete Fourier transform*, DFT [Proakis et. al. 1996], we can transform the sales from the time domain to the frequency domain. Then we can find the strongest low frequency in the data, which probably represents the basic seasonal behavior we are looking for. The DFT also gives us the phase for all frequencies. In the D3S2 module, the DFT is calculated with the *fast Fourier transform*, FFT [Singleton 1967] to increase performance. Using the FFT, finding the frequency of the seasonal variations in these relatively small datasets is very fast.

The seasonal trends cannot be expected to be perfectly sinusoidal. We can get a better waveform by using more frequencies, i.e. not only using one sine curve but two or more. In fact, if we use all frequencies in the DFT we can of course reconstruct the data perfectly. This is not what we want, however. The reason is that the reconstruction is periodic. If we use all frequencies to reproduce the data, the time period 0 to N, where N is the number of data points, will be perfectly reproduced but the period $N + 1$ to $2N$ will also be an exact copy of the first N weeks. Using this for predictions will simply produce the wrong results. A good compromise is to use only the strongest frequency detected and multiples of that. This means that we can use e.g. frequency *8*, *16*, *24* and so on. Using only multiples of the base frequency will give us a somewhat shaped waveform that is periodic in the base frequency, which is exactly what we want. When more complicated waveforms are necessary to make an accurate prediction, we can benefit from the fact that sales are strictly positive and that we might be able to represent the shape of the seasonal variation as a mixture of Gaussians, positioned using the EM algorithm [Holst, 1997].

When we have found the waveform, we need to find its amplitude. Most likely, this amplitude will follow the general trend of the data quite closely, so the best approach is to use a normalised seasonal component that is scaled with the general trend. The resulting model then consists of a base level, the general trend, with an additive waveform. This waveform is found using the DFT and possibly EM, and its amplitude is modulated by a line that also follows the seasonal trend. The model expresses sales as a function of time and media investments or brand tracking. Extrapolating this function into the future, the actual prediction for a specific period of time is calculated as the sum of sales given by the function during that time.

This model will produce predictions that are correct on average. This means that a request for a prediction of the sales during only one week or perhaps even one day will usually be off by a relatively large amount. The prediction gets better the longer time the user wants a prediction for, since it is rather accurate in average.

4. CONCLUSIONS

With the increasing demand for both short delivery times and less production for stock, sales forecasting for supply chain management will become more and more important. The work with the D3S2 module within the DAMASCOS suite has shown that by using all available information, it is possible to perform accurate sales forecasting and to provide information to the user that makes the forecasting a very usable tool that fits well into the normal workflow.

5. ACKNOWLEDGEMENTS

The authors would like to thank the EC for supporting the DAMASCOS project. They also would like to thank Nordisk Mediaanalys KB for sharing their expertise and knowledge on marketing data related issues.

6. REFERENCES

1. Cox R. T. (1946). Probability, frequency and reasonable expectation. *American Journal of Physics* **14**:1-13.
2. Holst, A. (1997). The use of a Bayesian neural network model for Classification Tasks. PhD thesis TRITA-NA-P9708, dept. of Numerical Analysis and Computing Science, Royal Institute of Technology, Stockholm, Sweden.
3. Jaynes E.T. (1986) Bayesian methods: General background. In Justice J. H. (ed.), *Maximum Entropy and Bayesian Methods in Applied Statistics*, pp. 1-25. Cambridge University Press, Cambridge, MA. Proc. of the fourth Maximum Entropy Workshop, Calgary, Canada, 1984.
4. Jaynes E.T (1996). *Probability Theory: The Logic of Science. Preprint: Washington University.* http://bayes.wustl.edu/etj/prob.html
5. Proakis J.G., Manolakis, D.G. (1996), *Digital Signal Processing – Principles, Algorithms, and Applications*. Prentice-Hall, New Jersey.
6. Singleton R.C. (1967), IEEE Transactions on Audio and Electroacoustics AU-15:91-97.

UTILIZING NEURAL NETWORK FOR MECHATRONICS, ON-LINE INSPECTION AND PROCESS CONTROL

Devdas Shetty[1], Noreffendy Tamaldin, Claudio Campana and Jun Kondo

College of Engineering
University of Hartford
West Hartford, CT 06117 USA
Tel: (860) 768-4615
Vernon D Roosa Chair Professor in Manufacturing Engineering
Shetty@hartford.edu

A trend in the area of automated manufacturing is the incorporation of artificial intelligence methods (example, neural network) to enhance the on-line inspection and process control. Intelligent on-line inspection and process control in modern manufacturing system have significant potential in improving production performance and quality of its product. Our research paper presents an approach for on-line inspection methodology that is applied to different areas such as metrology, precision measurement, mechatronics and bio-film thickness evaluation. In the case of metrology, the on-line measurement technique was applied for surface roughness measurement on the production floor. In the case of Mechatronics, the neural network technique was applied to the damped sensing and control system. Finally, an example of bio film thickness evaluation using neural network is presented.

Keywords: Neural Network, Mechatronics, On-line Inspection, Process Control.

1. INTRODUCTION

The advances in mechatronics and sensing have influenced immensely the field of inspection and process control in automated manufacturing. In conventional quality control system, robust design method and statistical process control methods are widely used. In robust design method, quality consideration is built-in during the design phase, even before the product is manufactured. (See Figure 1) As a contrast, statistical process control takes place at the inspection stage, where the machined product is being inspected and samples are taken for statistical data. In some industries such as microelectronics and aerospace industries, where constant monitoring of variables is needed, on-line monitoring is considered important. For on-line inspection and quality control, the neural network systems have the ability to monitor the production flow and take quick corrective action without taking the product out of the production line.

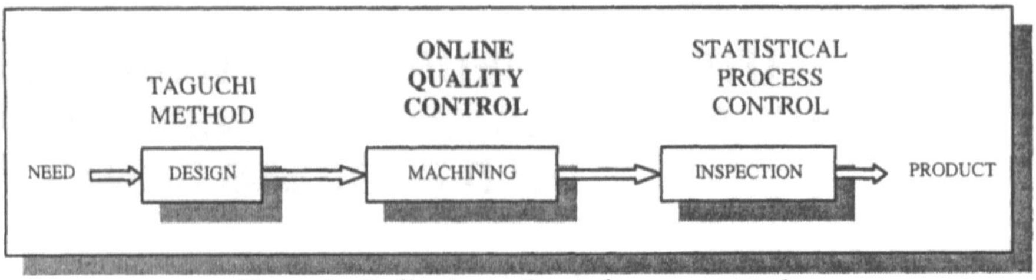

Figure 1 - On-line inspection and process control.

This paper highlights the application of neural network for three different situations. They are, (a) Vibration monitoring, and control using a mechatronic system (b) On-line surface roughness analysis (c) Neural network based pattern analysis.

2. LEARNING PROCESS IN NEURAL NETWORK

The idea of neural network comes from the biological system, the human's brain. The tiny brain cells, neurons perform millions of computations in a second so that human beings will have responses within our surrounding. A typical neural network system usually contains three main layers. The layers are input layer, hidden layer and output layer. Input and output layers usually contain nodes, which represent the input and output variables in the system. Hidden layer can be simply one layer and possibly more than one depending on the complexity of the applications. Every input has its own weight based on its contribution to the application and the weight. This determines the behaviors of a network. The typical neural network model is illustrated in Figure 2.

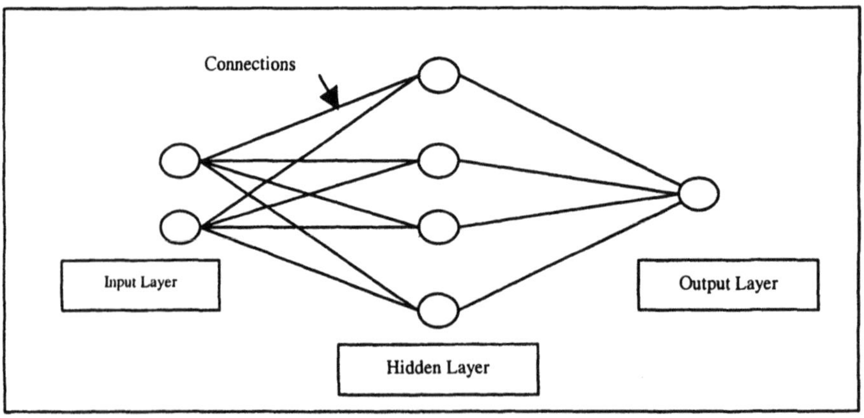

Figure 2 - Typical Neural network Model.

One of the important concepts of neural network is that it understands the learning process. In the learning process, the neural network can be itemized as supervised and unsupervised learning. Supervised learning is an important tool in neural network, and is based on the pattern of output produced by the network and the method of learning. For supervised learning, training is done by presenting a sequence of known input and output patterns. From that point, the system starts the decision making process while the operator could monitor and make any periodic changes in the training session.

3. VIBRATION MONITORING, DAMPING AND CONTROL USING MECHATRONICS

The principle of mechatronic demonstrator was researched for the purpose of monitoring the vibration by experimenting with various kinds of damping modes as well as control procedures. (Shetty et al., 1999)

The mechatronics demonstrator consists of three main assemblies such as mechanical assembly, sensing unit assembly and actuation system assembly. The mechanical assembly and sensing units have an arrangement of laser source where the light reflects from an oscillating mirror attached to a vibrating magnet. (Figure 3) The sensor collects the reflection and measures the displacement changes during the vibration phase. The oscillating magnet is held with a compliant spring system. This acts as a reflector and at the same time interacts with a voice coil actuator for controlling the damped oscillations.

Figure 3 - Sensing Unit Assembly with simulated laser beam.

The current flow to the voice coil actuator is controlled using hardware in the loop simulation. The damping of the system is designed using a model based control algorithm of the "rate feedback" type. The control algorithm is illustrated in Figure 4.

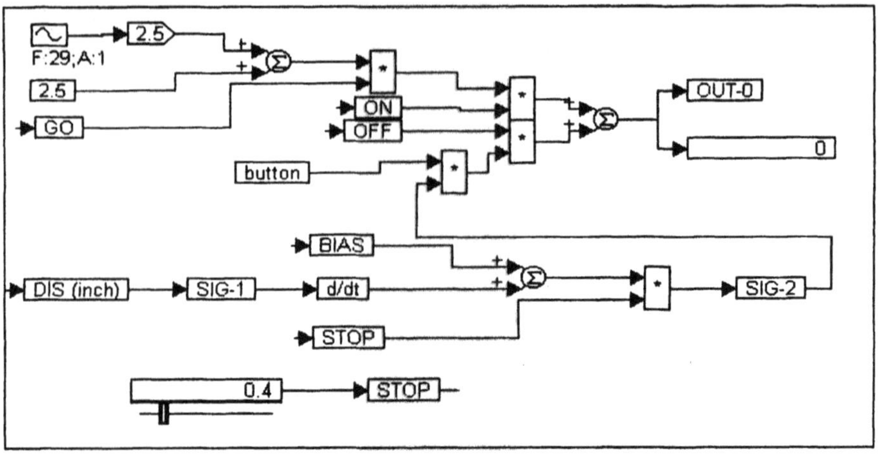

Figure 4 – Simulation Diagram for Rate Feedback.

The current supplied to the voice coil actuator and the displacement produced is monitored using the simulation diagram, and the reading is then compared to the ideal system, which is obtained through analysis and correlation developed based on the material properties of the vibration system such as, spring stiffness, damping factor, mass and other vibration related properties.

In real world applications, vibration is one of the factors that many engineer overlook. As a result of that, major catastrophes could happen and the structure built would collapse and perhaps cause life and financial losses. In the current case study, the experimental arrangement explores the natural frequency of its vibration. Maximum amplitude of the system is monitored through the position sensor, which differentiates the initial displacement and the final displacement. From the information obtained, we can establish a correlation between the forcing frequency supplied to the system and the maximum displacement produced as a result of that frequency. Further, a neural network program establishes the correlation between the forcing frequency and displacement of the target. This helps the designer to predict the displacement of the system without running the experiment.

The result of predicted amplitude or displacement (shown as theoretical) is compared to the actual amplitude and is shown in Figure 5. In this figure, we can clearly see that the actual amplitude did follow the shape of the theoretical amplitude with some offset value. One can obtain this by running the actual experiment in parallel with the prediction model simulated using visual simulation block diagram of second order differential equation. Both actual and prediction model produce a pattern called logarithm decrement and dies out in time. So with this information, we could easily control the vibration of the physical model so that it matches the predicted model as accurate as we want.

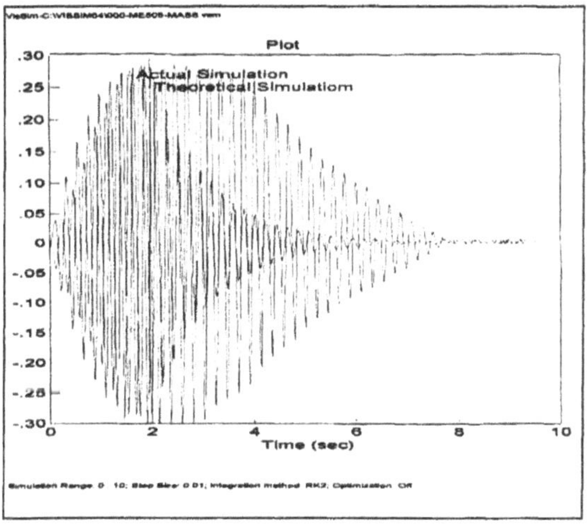

Figure 5 - Result of Predicted Model runs in parallel with Actual Model.

3.1 Neural Network Algorithm Dedicated to Mechatronic Applications

The algorithmic procedure explained in Figure 6 gives a better view of the whole procedure. The software developed is expected to predict the amplitude or displacement of the system based on the forcing frequency.

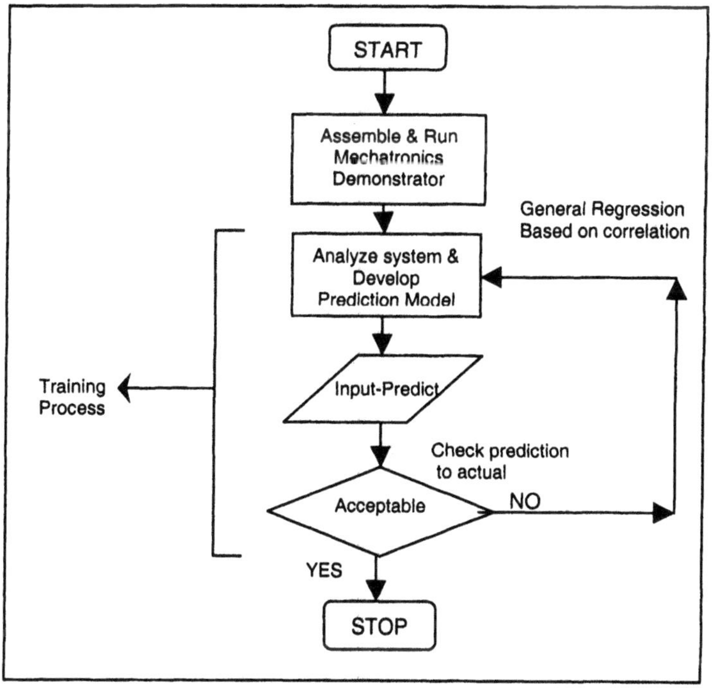

Figure 6 – Algorithmic Procedure

The graphical user interface of the custom built soft as shown in Figure 7 consists of three text-boxes and a command button. The three text-boxes are declared as frequency, amplitude and ratio. Frequency is defined as the input frequency (or forcing frequency) supplied to the system. Amplitude is defined as the predicted amplitude to the system as we supplied the input frequency (the final result). Ratio is defined as the resonance level of the mass spring damper system. This ratio is obtained by dividing the input frequency supplied to the natural frequency of the system (based on the mass and stiffness).

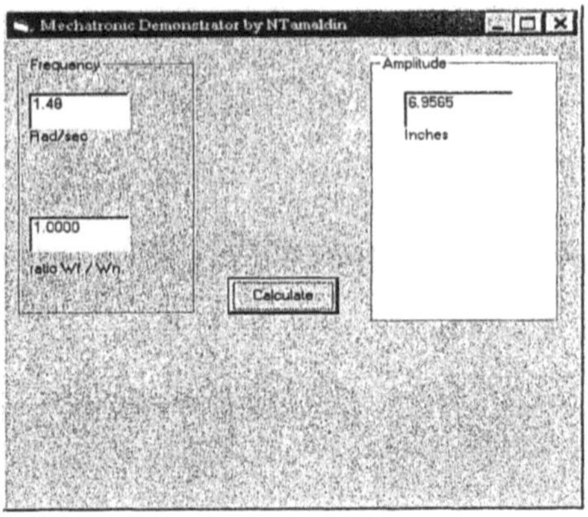

Figure 7 - Graphical User Interface of custom built Mechatronics Demonstrator.

The result obtained through the predicted system is then compared for verification with the physical model. In this part of the experiment, ten input frequencies are supplied to the custom built software and compared to ten actual reading on the system. As a result of that, the error analysis graph is plotted as shown in Figure 8 below.

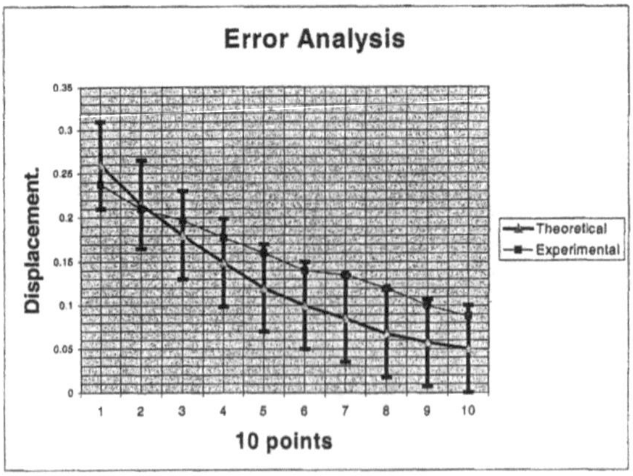

Figure 8 - The Predicted Value compared to the Experimental (Actual) Value.

From this plot we can clearly see that the experimental value did follow the pattern of the predicted value for most of the first ten runs. The error on point 7 and 8 is greater than the rest of the reading but it still lies within the 0.05 inches range of the predicted value. This error can be explained due to the noise and interferences to the experiment such as the detector unit, the electrical circuit and the voice coil actuator, which dissipate heat throughout the experiment through its coil as discussed earlier.

4. NEURAL NETWORK FOR SURFACE ROUGHNESS MEASUREMENT

The second example deals with neural network for the measurement of surface roughness of engineering surfaces using an optical scattering method (Embong et al., 1996). Light scattering has become a practical tool for measuring surface finish. It is a quick, sensitive, area sampling, and an inherently absolute method. The diffraction pattern from the surface consists of identifiable diffraction spots of maximum and minimum intensity of light as a function of angle. The determination of the frequency of diffraction spots leads to the information of the surface. The necessary link between scattering and surface topography can be made using either empirical correlation or an appropriate scattering theory. For implementation of the integrated intelligent system, this can be accomplished by feeding the appropriate input/output to the relationship.

An object and its farfield diffraction pattern have a Fourier Transformation relation with each other. If the object distribution is represented by $f(x, y)$, its Fourier transformation F(x,y) is given by,

$$F(x, y) = f(x,y)e^{-2p(xu+uv)}dydx$$

where x and y are spatial coordinates; u and v are spatial frequency variables.

The power spectral density function is an important function for expressing both the amplitudes and spacing and is identified by the diffraction pattern. The diffraction patterns for machine surfaces, resulting from laser light at different angles of incidence to the surface, are characteristic of that surface. Therefore, from this data, a determination of the type of machining performed on the surface can be made. The amount of dispersion in a diffraction pattern is directly related to the surface roughness of the material.

The objective of classifying the surface roughness of a machined workpiece is achieved by utilizing laser and a microcomputer-based sensing system. The intensity of the collimated monochromatic light source diffracted in the spectral direction is sensed by a number of photo-diode sensors that provides an analog signal to a digitizing system for conversion to digital information which is subsequently modified to display the surface roughness value. The intensity of the diffracted light is measured as a function of the output voltages produced by the sensors and is processed by the digitizing circuit. The method further includes a microcomputer-based procedure, which provides the operator interaction in the form of menu driven graphical interface. Based on different output voltage produced, an intelligent system, ***neural network's surface roughness identifier block*** estimates the roughness value of the specimen that is being inspected.

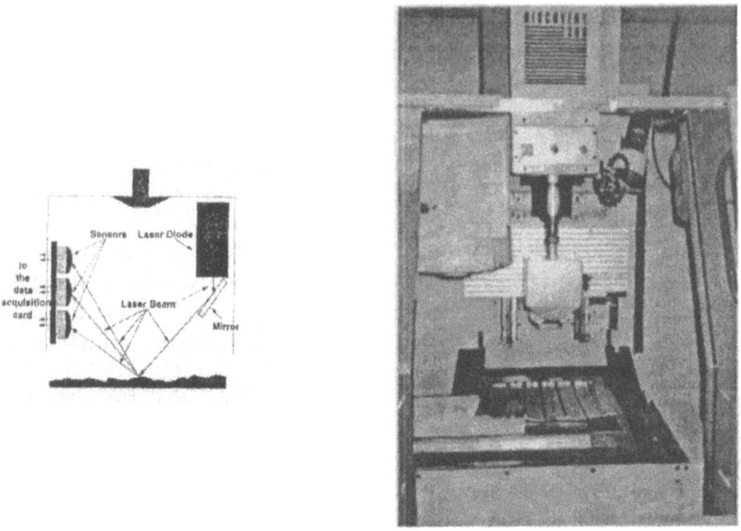

Figure 9 - Miniaturized Surface Roughness Analyzer CNC Tool Using Artificial Neural Networks

The experimental results of this prototype reveal that an intelligent integrated inspection system using neural network can reproduce surface roughness values to an accuracy of 99.5%. However, for various room conditions different from the learning process condition, the results should be analyzed for errors on account of differing operating conditions and fluctuations in input power. For on-line implementation, the miniaturized prototype of the intelligent surface roughness analyzer as shown in Figure 9 can be used (Hamdani et al., 1995). It offers small, compact, and rigid equipment to be mounted on the spindle of typical Computer-Numerically-Controlled (CNC) machine tool. This on-line device enables continuous monitoring of the surface roughness condition of the part being machined as the operator can readily switch the operation between machining and inspection process without having have to take the workpiece out of machining center. The displayed roughness value provides the necessary feedback and updates the operator on the current state of the machining process. The operator, thus, can adjust the machining process such that the surface roughness parameters will always be within the tolerance of the target value. This approach requires interactive communication between the CNC machine and the surface analyzer.

The methodology explained above integrates the miniaturized surface roughness analyzer with the CNC machine using a microcomputer-based sensing system. A solid-state relay system establishes a link between the CNC machine controller and the data acquisition system and provides two-way communication with the CNC machine.

The basic operation of the surface analyzer in an on-line mode consists of initial system set-up, calibration, measurement, and positioning of the machine tool. The machine tool used in this study is a vertical machining center, which uses the part programming language facilities. The inspection operation starts with the identification of the geometry of the workpiece, and (x, y, z) positional information of the points are determined- The surface analyzer, which is mounted on the spindle of the machining center has to be calibrated first with the standard roughness samples

Before the artificial neural network can be utilized, it has to go through learning procedures to establish the interconnection weight among the nodes within its layers. This can be done by supplying each of the sensor output into the identifier block as different node in the input layer, This information is transmitted to all the nodes in hidden layer, and eventually to the output layer, while producing relationship possibilities among the inputs and the output based on the actual output gained through the training input. Once all the critical values are identified, a standard bias transformation has to be done without affecting the ratio among each other since this ratio is essential to the identifier block in establishing its interconnection during the training process.

After the identifier is fully trained, the learning feature inside the artificial neural network's surface roughness identifier block is disabled and it is ready for predicting any specimen surface roughness value, Figure 10 shows the typical block diagram setting for learning process using visual simulation software.

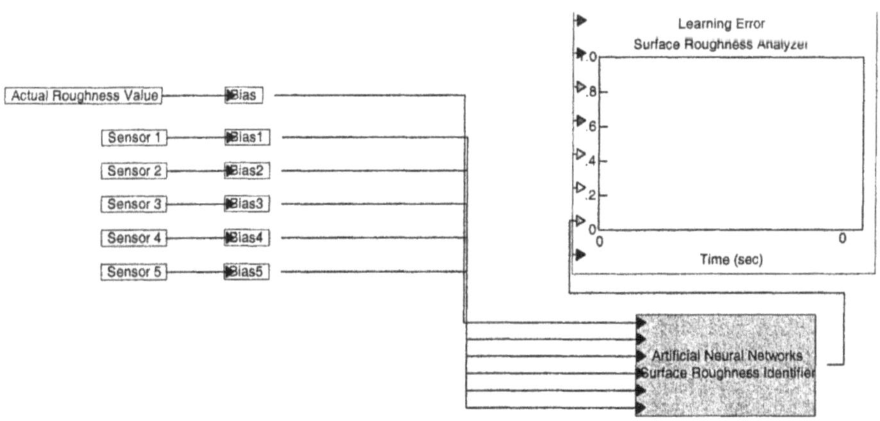

Figure 10 - Block Diagram Setting for Learning

5. NEURAL NETWORK FOR BACTERIA PATTERN EVALUATION

The third case study with neural network involves identification of samples of bacterial pattern. This has potential application for inspection of the inside surface of a large container or tank containing liquid medium without emptying the tank. This experiment is still in the early phase and needs more analysis and study. For the purpose of neural network application, the designer can create a database containing all the possible samples pictures taken with the surface roughness unit associated with it. The procedure adopted in the existing surface roughness machine can be adopted to measure the samples and save the pictures of the samples along with its surface roughness unit (Ra). The surface roughness machine is illustrated in the Figure 11 below.

Figure 11 - Surface Roughness Machine at the University of Hartford

Prior to taking the measurement of surface roughness from sample of bacteria, a correlation between bacteria growth and time is established. In this experiment, the designer assumes that the correlation has already been established and the relationship between bacteria growth is proportional with time until a point where the bacteria does not grow anymore. The plot of the assumption is illustrated in Figure 12. After 4 and half days, the bacteria stops growing and the thickness becomes constant starting from that point.

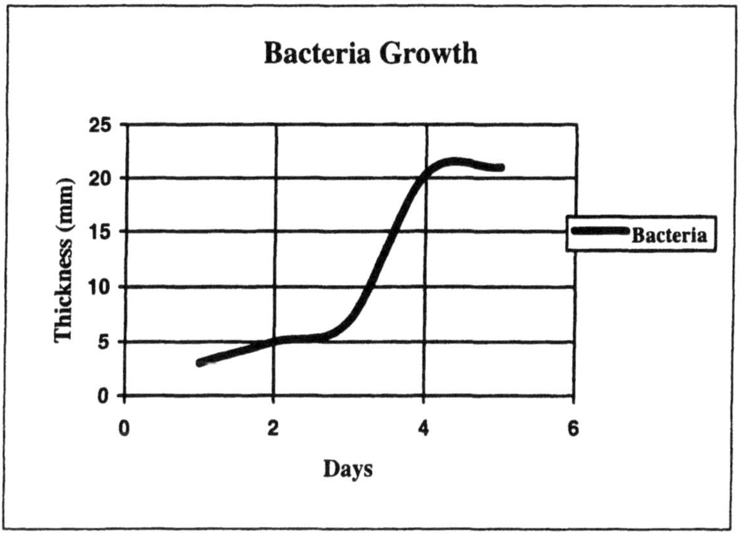

Figure 12 - Assuming this correlation is true for bacteria growth.

Once all the possible samples are taken and saved into the database, the neural network system is ready to perform the match for the possible new samples. Once the fully trained software is run, it should be able to find a match from the database and give the readings with the percentage of accuracy in every match that is found.

This method is arguable in terms of how many samples we have to take in order to complete the database. But, if we could measure the pattern and the thickness of the bacteria according to their growth, we might have more chances of successfully develop a complete database. Increasing the intensity of the laser beam aimed at the sample and intense the resulting pattern under microscope could also increase the precision of the surface roughness

6 SUMMARY

The three examples shown in this paper demonstrate how a real-time monitoring and inspection could be performed. The continuous monitoring of quality reduces the dependence on Taguchi and SPC methods which are obviously more costly and time consuming. Robust design requires careful pre-planning of integrating quality in the design process, while SPC is expensive mainly because the quality control is post-production and involves waste in form of discarding products that do not meet the customer specifications. On the other hand, the real-time method is less costly because it takes corrective action at the source of the problem in real-time, and it is a highly flexible quality control method because it uses neural network system. A neural network is very flexible because it continuously learns and makes quality control adjustments based on the data presented. In addition, neural network systems are not labor intensive. Once designed and trained, they can operate in an autonomous environment. The neural network area is growing in parallel to the computer industry since computers play a major role in the neural network methodology in process control and inspection. A large portion of production of consumer product is automated and with appropriate modification, the automated system could be improved with the utilization of neural network.

7 REFERENCES

1. Embong, S., Shetty, D., Motiwalla, L., Kondo, J., and Kathawala, Y., *Real Time Architechture for Advance Quality Monitoring in Manufacturing.* International Journal of Quality and Reliability, 1996
2. Hamdani, Z., Shetty, D., and D'souza, J., *Development of Online Surface Roughness Inspection,* University of Hartford, CT., May 1995.
3. Shetty, D and Kolk, R *"Mechatronic System Design"*, PWS Publications / Brooke Cole, Boston, USA, 1998
4. Shetty, D., Kolk, R., Kondo, J., Campana, *"Mechatronics Technology Demonstrator"*, University of Hartford, College of Engineering, 1999

SELECTION CRITERIA FOR MAKE OR BUY DECISION PROCESS IN ENTERPRISES

Beata Skowron-Grabowska
Management Faculty
The Technical University of Czestochowa
ul. Armii Krajowej 19b
42-200 Czestochowa, Poland
e-mail: beatas@zim.pcz.czest.pl

An enterprise establishes several selection criteria in material supply process. It is very important in decision processes to accept right selection criteria. These criteria are different in their nature. It is possible to divide criteria into two groups: quantitative criteria and qualitative criteria. On the base of quantitative criteria mathematical model was built. In this model material supply process is shown for such enterprise, which analyzes different sources of deliveries. Supply sources can be described by range of materials, kind of production, customer requirements, prices, size of demand. All above-mentioned reasons influence on the enterprise and force it to make decision to produce or to buy.

Keywords: make or buy, production, purchase, selection criteria, costs

Decision problems of make or buy can be considered in different aspects. The basic aspect follows from mathematical solutions. In the frame of this aspect vectors are used. These vectors describe quantities and prices of materials needed for production and at the same time these materials are the subject for make or buy decisions. Some vectors are assumed to describe materials produced in the enterprise and some vectors are for materials bought by the enterprises. In the first case it is received: vectors with quantity and costs of material production and in second vectors with quantity and prices of materials. As a consequence measurable criteria of make or buy selection are received.

Due to the complexity of the problem, making a choice between the production and the purchase of goods in enterprises requires methodological consideration. The aim of this is to formulate the requirements as to the method which will give a clear answer to the question: which is a better solution – to manufacture or to consider the purchasing of goods? The selection of criteria should be a necessary prerequisite in order to answer this question in a satisfactory manner. The criteria should be selected by the examination of the situation and the consequences of the already existing set of conditions.[i]

In the literature there are five rules verifying the accepted selection criteria.[ii]

- Aiming at the unequivocal character of the verification by reducing the number of criteria,
- Keeping a relatively comprehensive character of the evaluation by the exact description of the choice options,
- Clear definition of each criterion,
- Necessity of defining and measuring the consequences,
- Need of defining models for particular criteria.

Achieving a compromise between the first and the second criterion, allows of defining their optimum number.

It is also essential to define measurement methods. Taking into account the difficulty in describing the criteria, a unified evaluation system, either in the form of importance or another indicator, should be applied. The possibility of assessment of the criteria often aims at working out a model of behaviour towards a certain choice. Defining all criteria of evaluation is important as far as their hierarchy is concerned. In the economic practice the best criteria are the ones that have a normative character. In this case it is easy to evaluate whether the value of a given indicator stays within its norm or not.

The rules presented above confirm the importance of studies concerning the selection criteria. In the literature there is a great number of criteria adopted for verification. A significant number of criteria and their diverse character cause the necessity of their division. The carrying out of such a division aims for the systematization of the criteria from different points of view. One of them is assessment (Fig. 1). This criterion employs money and non – money factors. Within their range there are parameters whose qualification is difficult as well as the elements that are possible to measure.

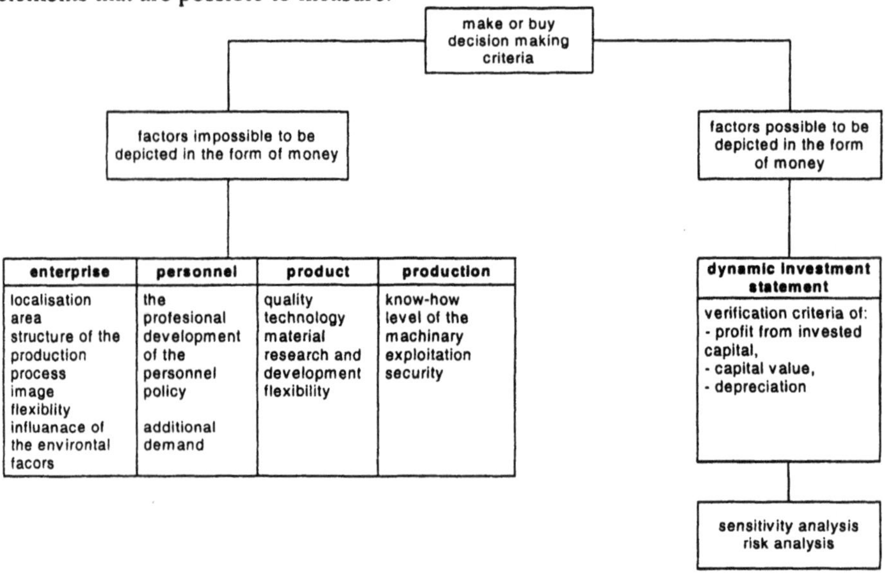

Figure 1 Selection criteria for make or buy decision process on the basis of money and non – money factors.

Source: Duck O., Schotz S., „Gospodarka materiałowa". Poradnik praktyczny Alfa – Weka Warszawa 1998 Chapter 2/10.2 p. 4

The basic division of the criteria concerning make or buy decision making is the following: the money criteria and all the others (qualitative, technological etc.) are distinguished. Such a division enables making the decision and, above all, the presentation of the results of studies after adopting such criteria.
There are three major ways by means of which the manufacture products necessary in a business activity may be delivered to the enterprise:

1. Manufacturing the necessary goods in the given enterprise,
2. Buying the manufacture products from suppliers or co-operating units,
3. Manufacturing the products and buying the missing elements from suppliers.

Another division of criteria that may be useful in the make or buy decision making process is presented below. The most important part of this division is the classification of the criteria into three separate groups, which enables an easier and faster decision making process. After being put into an appropriate class, in some cases the decision to be made is obvious; in many cases, however, it is necessary to carry out further studies and calculations. In accordance with the adopted rules there are three groups of criteria in the decision of whether to buy or to make the production goods. They are the following[iii]:

- Criteria with no choice of an alternative (there is no other choice than production). The most common cause of the choice of the production option are the restrictions concerning technology, licences and patents.
- Measurable criteria that employ the economic calculation and as a result it enables the optimum choice of the way in which a company is supplied with the production goods,
- A criterion that is difficult to measure or non-measurable at all and related to such categories as quality, time and flexibility of deliveries and one's own production. This criterion is also related to supply market, outlets, the company itself, etc.

There is a table presented below to enable a detailed analysis. It was described by J. Famielec, and it contains the basic criteria facilitating the make or buy decision making (Table 1).

Table 1 The basic criteria facilitating the make or buy decision in enterprises.

Make	Buy
1 first order criteria (with no choice of option)	
a. Necessity of keeping the production of given goods secret b. Lack of the possibility that they are bought from suppliers or co-operating units (the lack of them or no possibility of transport of the goods) c. Quality requirements (the lack of the fulfilling these requirements by a supplier	a. A law protecting the production of a given manufacturer b. Technological knowledge c. Shortages of one's own means of production d. Small needs e. Quality requirements impossible to fulfil by one's own production
2 second order criteria – economic calculation (choice)	

a. Lower costs b. Higher effectiveness	a. lower costs b. higher effectiveness
3 Third order criteria – difficult or impossible to measure	
Supply market	
a. protection from delivery disturbances caused by the market and extra – market factors b. gaining independence from suppliers and transport c. providing a high level of supply security in terms of supply deadline, quality and the cost of product	a. trade benefits b. the possibility of replenishing full supply range c. possibility of storing of the production goods by the suppliers
Outlet	
a. aiming at keeping monopoly as far as supplies and the final product is concerned b. diversification of the range of supply with semi-finished products and parts c. hindering the purchase of the product by competitors	a. taking part in the marketing system of suppliers b. diminishing the risk of the lack of demand for final products (sharing the responsibility with the suppliers)
Enterprise	
a. prestige b. flexibility c. exploitation of the processing power	a. diminishing the risk of the lack of technological advance and development b. observing the rules related to the signed contracts and agreements

Source: Famielec J. Wybór między wytwarzaniem a zakupem środków produkcji w strategii przedsiębiorstw przemysłowych, ZN AE w Krakowie, Kraków 1994 p.79

The juxtaposition of the arguments in favour of manufacturing or buying products or against those options allows their quick analysis and facilitates making a decision to select the supply source[iv].

The determination of the hierarchy of importance for the two options helps to choose the proper one. Thus, the enterprise reacts fast to the needs of the supply market as well as the outlet and is flexible to changes, by means of the close contact with the market. All of these factors lead to the increase in the competitiveness of the enterprise on the market, which is the main objective here.[v]

The second order criterion, i.e. the economic calculation based on costs, is easy to present in the form of a table. It is also easy to locate the source of costs connected with the process of supplying with the production goods. At each stage of the partial process there is an increase in costs.

Due to the comprehensive character of make or buy decision process, not only the cost criterion (connected both with the present and future costs), but also the quality criterion should be taken into account. In the quality criterion the following factors should be considered: economic capacity, the know-how production secret, technological innovations in the product or service, the quality of the manufacturing asset itself as well as production fluctuations (both qualitative and quantitative ones) arising from the factors on which the managers have or do not have any influence.[vi]

For a more detailed presentation of the qualitative criterion one may apply the procedure that analyses particular elements of the given criterion.

It is juxtaposed in the form of tables that allow a clear presentation of elements, which is essential when taking strategic decisions to which, undoubtedly, the make or buy decision belongs. /Table 2/

Table 2 A comparison of the possibilities of making and buying with the use of the qualitative criterion.

Qualitative criterion	Weight	Own production		Supplier	
		Points	Points with weights	Points	Points with weights
C_1	w_1	p_1	$w_1 p_1$	d_1	$w_1 d_1$
C_2	w_2	p_2	$w_2 p_2$	d_2	$w_2 d_2$
C_3	w_3	p_3	$w_3 p_3$	d_3	$w_3 d_3$
C_n	w_n	p_n	$w_n p_n$	d_n	$w_n d_n$
Total	1.0	Σp_n	$\Sigma w_n p_n$	Σd_n	$\Sigma w_n d_n$

Source: the author's own analysis, based on: Krawczyk S., „Logistyka w zarządzaniu marketingowym" Wyd. AE Wrocław 1998 p.149

While drawing up the table above, first of all, the hierarchy of importance for particular criteria should be determined. Depending on the priority adopted for each element, appropriate significance should be given under the following conditions:

$w_n \geq 0$ and $n = 1,2,3,\ldots,N$

and

$$\sum_{n=1}^{N} w_n = 1$$

It is important for practical reasons. Such a calculation algorithm allows a better clarity. One should set up punctual quantities, p_n for the enterprise and d_n for the supplier. The range is determined by the most and least beneficial parameters for each element.

The adopted hierarchies of importance w_n may be corrected according to the changing needs and priorities established in the enterprise. The comparison of hierarchies of importance and points in one table allows a quick analysis and the numerical data concisely present the quality criterion results and compare them with the cost criterion.

It should be stressed that in the process of criteria selection the main role is played by an employee.[vii] It is a decisive factor in the make or buy decision process and it is their experience and analytical skills that the right decision depends on.

It is essential, therefore, to consider pointing out such an alternative of material supply that is profitable for the enterprise. The potential come from economic calculation model solutions. It allows for a choice of supply alternatives.

In these calculations it is assumed that the basis for the analysis is space A* within which alternative solutions are searched for.

Space A* is closely connected with the enterprise and its environment as the set *k* of all possible kinds of acceptable supply alternatives. Within the range of choice there is a source of supply, the kind of material, its quality parameters, unit price, etc.

If all possible alternatives of supply of an i-kind are designated by A_i^* (where i = 1,2,...,k), the space A* is the multiplicity sum of the sets $A_1^*, A_2^*, \ldots, A_k^*$, which means:

$$A^* = A_1^* \cup A_2^* \cup \ldots \cup A_k^*$$

From space A* a subset A is isolated, by indicating the subsets $A_1, A_2, \ldots, A_k$ of the sets $A_1^*, A_2^*, \ldots, A_k^*$ respectively. The above subsets contain more detailed characteristics of the material supply, for example the supply conditions, the co-operation level.

Thus, there is

$$A_i \subset A_i^* \qquad i = 1,2,\ldots,k$$

and

$$A = \bigcup_{i=1}^{k} A_i \subset \bigcup_{i=1}^{k} A_i^* = A^* ,$$

so

$$A \subset A^*.$$

Set A contains alternatives of choice concerning the enterprise and set A* - all acceptable alternatives of choice.

Both the cause and criterion of such an isolation is the fact that not all alternatives of supply choice acceptable in space A* currently occur in the enterprise, i.e. they are available at a definite time.

It is assumed that A is a subspace of space A* containing only those alternatives of material supply that are indispensable in the activity of the enterprise. It means that subspace A contains all possible alternatives of choice of material supply of the *i*-kind (*i* =1,2,...,k), that exist. Subspace A contains alternatives of material supply choice. These are the solutions depending on one's own production, buying the materials or services from suppliers. Thus, certain relations develop in the spaces in question, where the above-mentioned solutions are juxtaposed. The interrelations between the sets mentioned above are depicted in Fig 2.

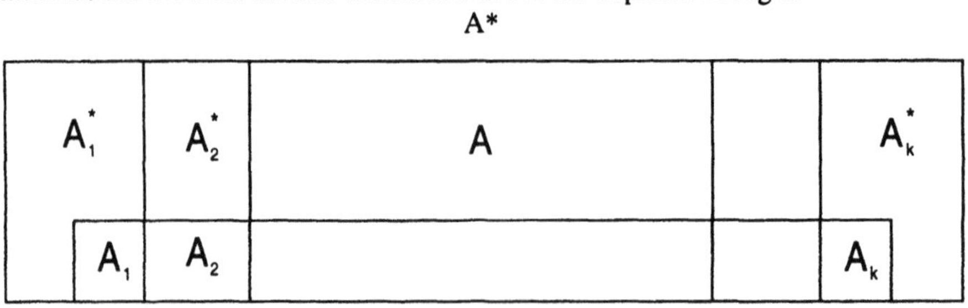

Figure 2 Relations between sets of alternatives of make or buy choice.

Source: the author's own analysis

$$A_i \subset A_i^*, \quad i = 1,2,...,k$$

$$A^* = \sum_{i=1}^{k} A_i^* \qquad A = \sum_{i=1}^{k} A_i$$

Set A_i for $i = 1,2,...,k$ of materials of i-kind is divided on the multiplicity sum of the two following sets: P_i – products of the i-kind manufactured by the enterprise and the set Z_i – products of the i-kind manufactured (or provided in the case of services) by other enterprises, thereby bought by the enterprise in question.

In this case it may be stated that

$$A_i = P_i \cup Z_i \qquad \text{for } i=1,2,...,k$$

$$A = \bigcup_{i=1}^{k} P_i \cup Z_i = \bigcup_{i=1}^{k} P_i \cup \bigcup_{i=1}^{k} Z_i,$$

because

$$P_i \cap Z_i = \phi \qquad \text{for } i=1,2,...,k.$$

The equation presented above means that sets P_i and Z_i are disjoint sets, i.e. their common part is an empty set. Each unit of the i-kind material belongs thus to set A_i, (which means it belongs to set $P_i \cup Z_i$), but does not simultaneously belong to the sets P_i and Z_i. The enterprise takes the decision then: either it makes or buys products or semi – finished products.

Thus, on the basis of the model presented above several alternative possibilities may be considered.

References:

1. Cabala P., Obiektywizacja preferencji istotności kryteriów oceny projektów gospodarczych, ZN nr 509 AE w Krakowie, Kraków 1998
2. Duck O., Schotz S., Gospodarka materiałowa. Poradnik praktyczny, Alfa – Weka, Warszawa 1998
3. Famielec J., Wybór między wytwarzaniem a zakupem środków produkcji w strategii przedsiębiorstw przemysłowych, ZN AE w Krakowie, Kraków 1994

4. Krawczyk S., Logistyka w zarządzaniu marketingowym, Wyd. AE we Wrocławiu, Wrocław 1998
5. Lewandowska L., Alternatywne formy finansowania w procesie zarządzania logistycznego, conference materials Przedsiębiorstwo na rynku kapitałowym, Uniwersytet Łódzki 2001
6. Rayburn L.G., Cost accounting. Using a cost management approach, IRWIN, Boston 1993

---

[i] Cabała P., „Obiektywizacja preferencji istotności kryteriów oceny projektów gospodarczych" ZN nr 509 AE Kraków „Prace z zakresu procesu zarządzania" 1998 p. 79
[ii] Cabała P., „Obiektywizacja preferencji istotności kryteriów oceny projektów gospodarczych" ZN nr 509 AE Kraków „Prace z zakresu procesu zarządzania" 1998 p. 80
[iii] Famielec J., „Wybór między wytwarzaniem a zakupem środków produkcji w strategii przedsiębiorstw przemysłowych" Kraków 1994 Zeszyty Naukowe AE p.78
[iv] Famielec J., Wybór między wytwarzaniem a zakupem środków produkcji w strategii przedsiębiorstw przemysłowych, ZN AE, Kraków 1994, p. 80
[v] Lewandowska L., „Alternatywne formy finansowania w procesie zarządzania logistycznego" conference materials „Przedsiębiorstwo na rynku kapitałowym" Uniwersytet Łódzki 2001 p.456
[vi] Rayburn L.G. „Cost accounting. Using a cost management approach" IRWIN Boston USA 1993 p.564
[vii] Rayburn L.G., „Cost accounting. Using a cost management approach" IRWIN Boston USA 1993 p.564

DESCRIPTION OF THE COMPETENCE CELL PROCESS PLANNING

Holger DÜRR [1], Jens MEHNERT [2]
[1] Prof. Dr.-Ing. habil.
Head of Department of Manufacturing Technology, TU-Chemnitz
E-mail: *h.duerr@mb2.tu-chemnitz.de*
[2] Dipl.-Ing.
Member of Department of Manufacturing Technology, TU-Chemnitz
E-mail: *jens.mehnert@mbv.tu-chemnitz.de*,

The advantages of the non-hierarchical networking model are the elimination of the dominance of single companies and the creation of more intelligibility regarding yield and customer communication within the network. This is the basis of customer-specific cooperation of competence cells formed by existing enterprises. One result is the reduction of costs for administration and coordination. When realizing the demands of the non-hierarchical network model and at the same time considering the structure of the competence cell therein, questions for necessary development of the competence cell process planning arise for this form of cooperation. How should this competence cell be structured? With which methods should production courses be organised to meet the requirements of this special model of distributed added value?

Keywords: Process Planning, Configuration of Production Networks, Description of Competence Cells, Competence Presentation, Small and Medium Sized Enterprises

1. VISION OF A NON-HIERARCHICAL PRODUCTION NETWORK

Present cooperation forms rely on hierarchical structures in and between enterprises. Small and medium sized enterprises, for example, often take a big and economically powerful enterprise (e.g. from the automotive industry) as a partner with whom they form a relatively stable and long-term cooperation relation (e.g. for a series). This cooperation relation is dominated by the large scale enterprise in a mainly technical organisational and economic way. (Uhlmann, 2000) The gained advantage of relative stability is reduced by the disadvantage of one-sided dependency and the absence of immediate contact to the ultimate consumer.

These dependencies are restrictive towards regionally established enterprises and form considerable market-entrance barriers for newly-founded innovative enterprises. This leads to incomplete utilisation of regional potentials and competences as well as to the obstruction of economically-desired dynamics in the foundation and development of small enterprises. At this point the Collaborative Research Centre (CRC) 457 "Non-hierarchical Regional Production Networks" of the University of Technology in Chemnitz tries to offer solutions (SFB 457, 2000). The concept of CRC 457 takes the approach of linked, self-organised and equal competence cells as a basis in order to achieve the economically-desired and necessary effects described above.

In the vision of this network model, enterprises separate into independent viable competence cells which then form the elementary units for a customer-oriented, cooperative and non-hierarchical added value. Therefore, the aim is to modularise the sequences of the value-adding process with regard to this innovative cooperation form and to decompose single competences in a way that autonomous, cross-linkable, function-oriented and smallest value-adding units can develop (Figure 1).

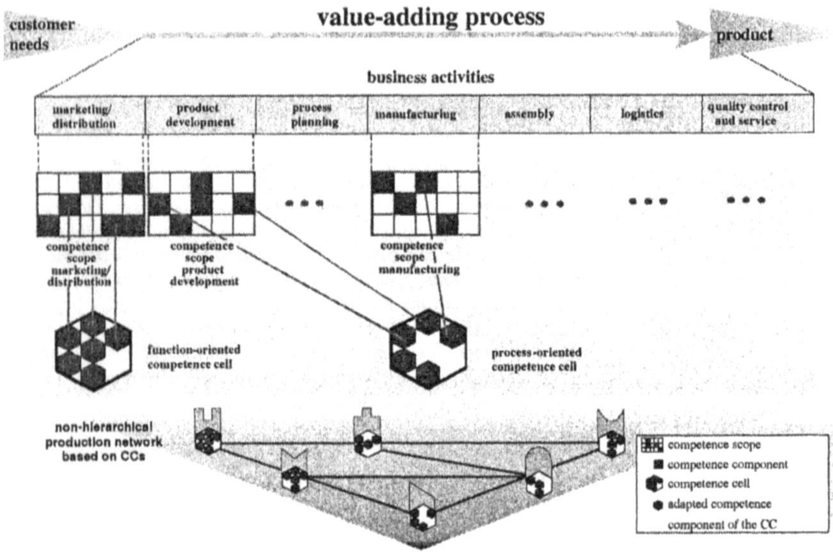

Figure 1 – Connection between added value process, business process, competence frame, competence cell and network

Considering customer-specific demands, these competence cells form production networks that are aligned to the best-possible achievement of customer orders and that are able to react very dynamically on the changing demands of the market. In order to achieve this, it can be necessary to organise competences from different sequences of the value-adding process together in one competence cell. Sometimes this is only temporarily necessary. The following text shall explain especially the demands and the possible developments of the competence cell process planning within the non-hierarchical regional production network.

2. DEMANDS ON PROCESS PLANNING IN NON-HIERARCHICAL NETWORKS

The non-hierarchical network model is characterised by a number of qualities which show the contrast to other existing cooperation forms. These characteristics form the basis of an order-oriented formation of competence cells and of value-adding processes for a distributed manufacturing in the organisation form. The factors of influence on the design of the containing process planning competence represent a subset of the whole target system. The in the following described qualities of this cooperation form are important for the modelling of the competence cell process planning.

Table 1 shows the most important factors of influence and their impact on process planning. Knowledge of which is a prerequisite for modelling the partial model of the process planning competence cell. At the same time, these demands are an essential part for the order-oriented definition of internal and external interfaces as well as for the representation of potentials of process planning competence cells within the innovative network model. In the following the planning methods to be applied are oriented at it (Dürr, 2001).

Table 1 – Factors of Influence and Demands on Process Planning

Factor of Influence	Demands on Process Planning Competence
Non-hierarchical organisation of performance creation	Provision of methods, models and instruments for planning and operation of networks. Establishment of substructures for self-organisation
Added value through smallest performance units (Competence Cells)	Customer-oriented and direct networking of CCs and utilisation of numerous possibilities of composition of CCs
Ways of manufacturing limited to individual part and small series fabrication	Focus on these ways of manufacture during the creation of corresponding planning methods and models
Customer-determined spectrum of products	Support of customer-oriented production planning by providing suitable planning methods
Starting point of CC are small and medium sized enterprises	Adaptation of planning conditions and planning methods to this enterprise category
Highly dynamic and mostly temporary structure of the network model	Development of flexible and dynamic competence cell structures and appropriate process planning methods
Necessity of transforming semantic information contents into machine-processable form	Stringent representation of process planning processes including their interactions in an ontology-based database in the ITCM

Likewise the evaluation methods, deposited in the information-technological core of the model (ITCM), are based on these demands. The evaluation methods are used in search for suitable process planning competence cells and for the following selection of those prioritised for the distributed added value (Neubert, 2001).

3. STRUCTURAL MODEL OF THE PROCESS PLANNING COMPETENCE CELLS

In order to identify a solution, the added value process is decomposed from the perspective of process planning. The developed activity diagram (Figure 2) shows the planning procedure of the manufacture of components for mechanical engineering in general. The usage of Unified Modelling Language (UML) (Fowler, 2000) for notation supports the definition of internal and external interfaces of CC as well as the exchanged information in connection to their temporal reference to added value.

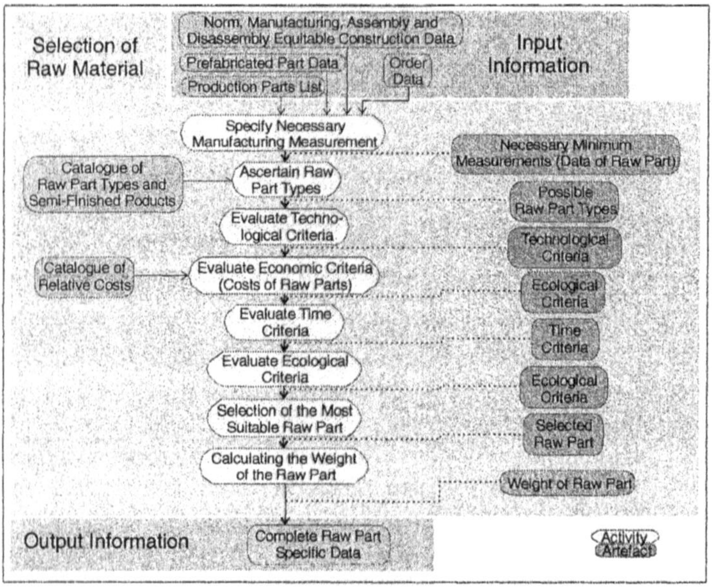

Figure 2 - The Developed Activity Diagram

The combination of these cognitions with the favoured operator model based on the Extended Value Chain Management concept (Teich, 2001) allows the demand-oriented conflation of the relevant content to the structural model of process planning competence (Figure 3). The partial models, which are currently shaped by engineer-technical understanding of this competence, are to be completed with important business and social components. Therefore only the first step towards the complete representation of process planning competence in the network has been successfully developed which means the engineer-technical design. The reference of process planning competence to product, performance and method is being used for the representation of it. Subject of further analysis are the necessary social and business aspects for a final selection of a definite competence cell.

The collection of the product data generated in the design and construction process takes place in the product model. There, the data relevant for the construction process are systemized and represented in a suitable way. The process planning competence cell uses, among other things, the functionalities of the product model for the representation of its performance abilities in the ITCM. Furthermore it defines itself about its ability to design materials, geometry characteristics and quantity-dependent knowledge of manufacturing and planning. This makes a product-related description of or search for process planning competence possible. The external interface design of the competence cell process planning is based in terms of the product model (e.g. at geometry characteristics) on a component classifier (Greska, 1995), while the material description makes use of the notation of the generalisation for the representation of material classes.

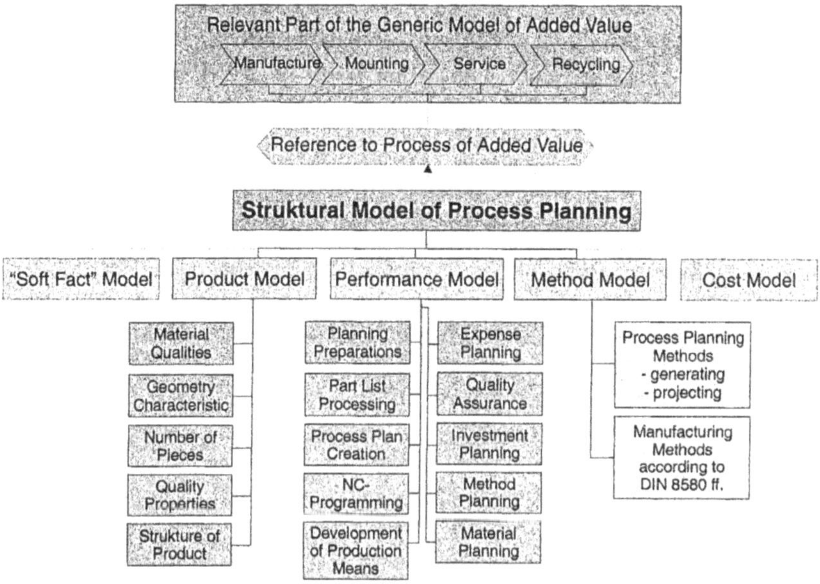

Figure 3 - Structual Model of the Process Planning Competence Cell

The method model reproduces process planning and manufacturing methods. The factors of influence of the non-hierarchical networking model will frequently lead to a combination of process planning methods. The new planning of process chains (i.e. order-neutral coarse planning) is used very often because of the small share of recurring same or similar components. The production tool specific detailing frequently allows projecting process planning methods (i.e. order-related detailed planning) on the basis of defined manufacturing form elements. The form element classes, that are aggregated in the component classifier, ensure the necessary semantics to the ITCM and support the external presentation of the description or the search of competence-cell-specific planning knowledge.

The manufacturing methods are based on the manufacturing processes in DIN 8580 ff (Deutsches Institut). The choice of this norm ensures on the one side the semantics for manufacturing competence and is on the other side a widespread external interface for description and search. An expansion by manufacturing methods which do not contain this norm takes place following the used description algorithms and ensures thereby imaginable supplementations or restrictions. This model is being used by the process planning competence cell as well as by the manufacturing competence cell. Beyond that it serves as a basis for the request vector in search for suitable manufacturing competence cells.

The performance model represents directly a link to the generic model of added value. The model contains the result of the decomposition - the universally valid activities of the process planning process in mechanical engineering. These activities have two tasks. Firstly they are the basis of the knowledge-base for planning manufacturing scenarios in the network. Secondly they are used for the detailing of descriptions of existing or needed process planning competence by the ITCM. The external representation of the performance model takes place in the generic model of added value within the authority-frame process planning.

4. TASKS OF PROCESS PLANNING IN THE NON-HIERARCHICAL PRODUCTION NETWORK

Embedded in the network identity of non-hierarchical production networks, the competence cell process planning has essentially to solve six tasks. These are:

- The generation of variants of possible manufacturing process chains as a basis for the subsequent order-related network configuration.
- The evaluation of the generated manufacturing process chains with the targets of the customer in mind.
- As an autonomous integral part of the network being the service provider for the potential customer of the network.
- Supplying process planning competence to the optimisation of production in the competence cells of the network.
- Serving as a link between marketing and construction for the calculation of offers in the network.
- Backup and maintenance of the descriptions of manufacturing competence cells stored in the ontology-based database.

The number of the just mentioned tasks and the special conditions of the non-hierarchical networking model cause the necessity of an individual adaption of the competence cell process planning. This leads to different characteristics of process planning competences within the structural model of the represented partial models. In the further process these new demands on the process planning process can only be met by modification and further development of the already mentioned process planning activities. In the following a special focus will be on the changes in the approach to the creation of the process plan.

5. EXAMPLE OF A PROCESS PLANNING CC

Until recently enterprises favoured the sequential methodology of process planning. However, taking the demands on process planning and the characteristic identity of the non-hierarchical network model as a starting point, it is recommended to use a less formal approach towards the creation of workflows which are not based on the direction of a superior control and monitoring instance. This mainly causes the remake of the function of the process plan creation. Starting with the selection of raw pieces, a technology-strategy-determination already takes place. Within the planning process this allows an early feedback with the customer.

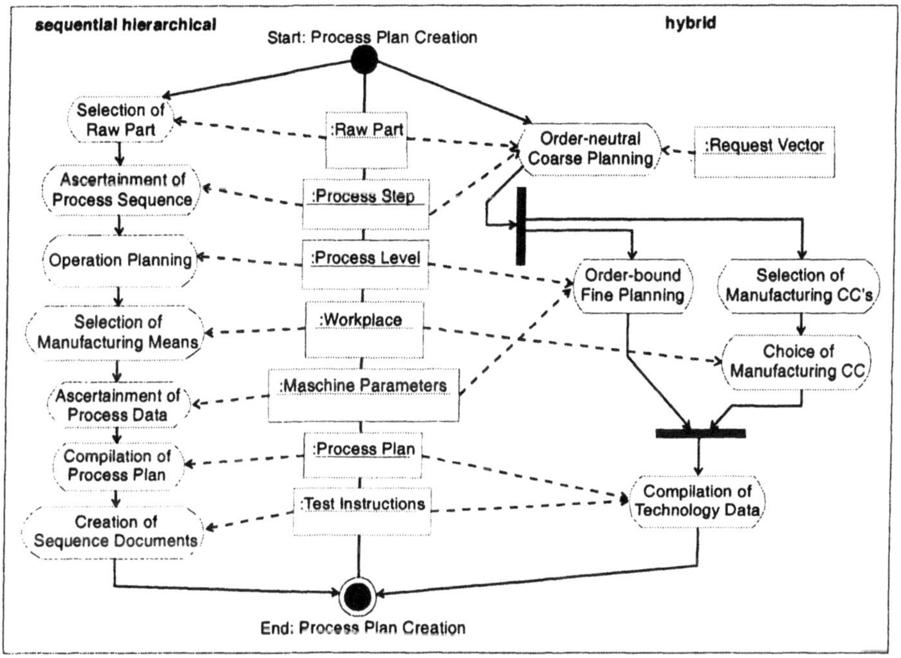

Figure 4 – Strategies of Process Plan Creation

The feedback points out manufacturing-method-related differences of the blanks with their specific influence on the desired product and the differing manufacturing parameter (purely quantitative) like delivery times, costs or processing times. At this point in time the ITCM with its managed functionalities serves for information procurement, e.g. for aimed research for electronic market-places. After the customer-oriented selection of blanks, order-neutral process steps are generated in order to obtain the geometry of the prefabricated parts. This happens in terms of coarse planning regardless of available or obtainable means of labour. The manufacturing method, that is represented in a process step, is being completed with information concerning form changes and evaluated. The result is a request vector for the search for corresponding manufacturing competence. This vector does not contain any data concerning costs and times of the manufacture. This makes the creation of numerous request vectors for the changes of geometry possible with defensible expenses. Thus the space for decision for the most favourable realisation of the network contains many options.

The selection of manufacturing means takes place through the adjustment of request and description vector of the manufacturing competence cell. This competence cell performs a resource-related detailed planning according to the requested processing task and completes individually the respective request vector with the relevant data. The advantage of this separation in order-neutral coarse planning (process planning) and order-related detailed planning consists of the possibility of an early machine data determination because this determination can be specified and stored order-neutral for complying production-orientated form elements. Likewise a central and order-related determination of these specific values cannot take place because of the expected spectrum of available means of labour.

The same is valid for the determination of process data. These considerations lead to the process planning competence cell taking on the task of creating order-neutral process variants for the distributed manufacture, which are used for the search for the necessary manufacturing competence. The described changes in the creation of the process plan reflect the high demands of this network model, especially with regard to a non-hierarchical and dynamic configuration of the manufacturing competence cell. The request-related linkage of generating process planning methods for the technology-strategy-determination with the projecting process planning methods for the process-data-determination leads to hybrid process planning methods. That means that the competence cell process planning makes use of special hybrid process planning methods in order to solve the set task.

6. OUTLOOK

Taking the structural model of process planning as a foundation, a submodel emerges for this added value sequence which is enhanced by the function view.

The generalisation of the characteristics of a competence cell is represented in direct relation to the added value process and the existing knowledge. The activities described in the performance model serve with their associations to the methodology and product model as a basis for the following creation of competence components. The knowledge necessary for the realization of activities gives information about the qualification of humans in a competence cell. Beyond that it also allows to draw conclusions concerning life cycle and development possibilities of competence cells.

7. REFERENCES

1. Deutsches Institut für Normung e.V., Berlin, Beuth Verlag GmbH, DIN 8580 ff
2. Dürr, H., Mehnert, J., Process Planning in Non-Hierarchical Production Networks, Proceedings of FAIM 2001, Dublin, ISBN 1-872-32734-6.
3. Fowler, M., Scott, K., UML konzentriert, Addison-Weysley Verlag, München 2000, ISBN 3-8273-1617-0
4. Greska, W., Wissensbasierte Analyse und Klassifizierung von Blechteilen, Carl Hanser Verlag München, 1995, ISBN 3-446-18462-7
5. Neubert, Ralf; Görlitz, Otmar; Mehnert, Jens: IT Unterstützung der Genese von Fertigungsnetzen. In: Hierarchielose Regionale Produktionsnetzwerke. Teich, Tobias (Hrsg.). GUC-Verlag Chemnitz. 2001. ISBN 3-934235-18-2
6. SFB 457: Finanzierungsantrag „Hierarchielose regionale Produktionsnetze" der TU Chemnitz für die Jahre 2000, 2001, 2002: http://www.tu-chemnitz.de/sfb457/
7. Teich, T.; Dürr, H.; Mehnert, J.: Prozessplanung innerhalb des Extended Value Chain Management (EVCM) Betreibermodells. Industrie-Management Heft 3/2001. Gito-Verlag Berlin. ISSN 1434-2308
8. Uhlmann, E.; Kröning, R.; Schröder, C.: „Regionale Kooperationsplattform" Form + Werkzeug, September 2000; S. 10 – 13

AN INNOVATIVE ICT PLATFORM FOR DYNAMIC VIRTUAL ENTERPRISES

Giordana Bonini, Vania Bicocchi
Democenter, viale Virgilio 55, 41100 Modena, Italy
Tel +39 059 899611 Fax +39 059 848630
Email: g.bonini@democenter.it
v.bicocchi@democenter.it

Flavio Bonfatti, Paola Daniela Monari
Dept. Of Engineering Science University of Modena and Reggio Emilia (Modena), Italy
Tel +39 059 2056138 Fax +39 059 2056129
Email: bonfatti@unimo.it

This paper aims to present the results achieved within the framework of the EU funded IST–1999–10768 BIDSAVER (Business Integrator Dynamic Support Agents for Virtual Enterprise) project. The project identifies an independent entity, the Business Integrator (BI), which co-ordinates and facilitates the path of the Virtual Enterprise. Specific software requirements from the BI point of view and from the SMEs point of view have been identified and used to select and tune the final innovative ICT platform used and validated during the project (that is one of the main project outcomes).

Virtual enterprises, ICT tools, requirements, architecture, pilot results

1. INTRODUCTION

1.1 The context

The concepts of Virtual Enterprise, holonic systems, distributed enterprise, multi-site and multi-business organisation, network organisation, and the like, are becoming increasingly important at national and international levels, and the need of new methodologies and planning tools for the SMEs is even more urgent. New global market opportunities and the development of innovative products require financial and technical capability that individual SMEs can provide with difficulties (Hirsch et al., 1995). The challenge facing organisations that want to form "virtual" partnerships is not in building of the actual partnership but in using it to its full potential (Filos et al., 2000). Companies encounter a lot of difficulties and problems in trying to set up new VEs because of several important factors like the lack of resources, poor documentation on past experiences, environmental and cultural factors that are heavily dependent on the classic concept of enterprise. Finally the biggest obstacle is the lack of proper and easy to integrate ICT tools to support the SMEs during the constitution and operation of the VE (Bonfatti et al., 1996).

1.2 The BIDSAVER project

Here we intend to present the results achieved in the field during the conduction of the project IST-1999-10768 BIDSAVER (Business Integrator Dynamic Support Agents for Virtual Enterprise). BIDSAVER developed a methodological, technological and legal framework to support competitive potential of SMEs through both the constitution and operation of Virtual Enterprises (VE), joining geographically dispersed SMEs in order to seizes business opportunities, and is managed, in order to enhance their competitiveness by a new entity, the Business Integrator (IST-1999-10768, 1999). BIDSAVER delivered an ICT prototype for the Virtual Enterprise information system, leveraging on e-Commerce and Internet technology. The prototype is based on pre-selected commercial solutions than can be configured according to VE requirements and improved through add-ons accommodating specific functions necessary to the success of the VE operation. The system is constituted by a number of commercial applications and three main management modules:

- a **Business Breakdown Structure management module** (BBS), allowing the management and evolution of a VE, both during the constitution phase (where the breakdown structure evolves according to the increasing detail in product definition and to the consolidation of the partnership structure) and during the operation phase
- an **information capturing agent** (ICA), performing the actual search for potential partners and providing updated information on co-operation opportunities during VE operation.
- the **Business information Integration module** (BIIM), in charge of ensuring integrated management of functions needed for the VE operation.

During the project, the provided ICT prototype has been applied to two real industrial cases (to test, improve and validate it):

- a network pilot case, focusing on the development of micro-satellites
- a supply chain pilot case, operating in the mechanical field

The Supply chain VE is a Virtual Enterprise working in the field of agricultural machine building. During the project, Democenter played the Business Integrator role for the Supply Chain: an experiment phase (that lasts 15 months) to test and validate the provided tools has been planned and carried out, supervised by Democenter. The paper aims at presenting the specific and innovative IT solution selected, adapted and developed in the framework of BIDSAVER project: the modules of the final ICT platform and the features that conducted to select the described tools are detailed in the paper together with the context of the Virtual Enterprise requirements to be supported by the platform. The paper is organised into 5 main sections: the Introduction (this section), a second section describing the characteristics of the Supply chain pilot used to select to tools. Section 3 provides a detailed description of the BIDSAVER platform (for each module). Considerations, parameters and the final opinion of the users involved in the pilot are reported in the Conclusions. The last chapter contains the references used in the paper.

2. THE SUPPLY CHAIN PILOT CASE

The Mechanical Equipment Supply Chain is a Virtual Enterprise formed by SMEs operating from geographically remote locations, that co-operate to give the external customer complete and operative products and services in the mechanical field. The Supply Chain works mainly in the production of components for agricultural machines and it is formed by COMER (VE Leader) and two partners: Livarna from Slovenia and Intermec from Slovenia.

To test and validate the ICT platform provided by BIDSAVER project, a specific product of the network has been selected: the group of gear-box for soil tillage machines. The main components of the product are (Deliverable D1, 2000):

- A 'BOX' that is the external structure (of the product) generally made of aluminium alloys and cast iron with several shapes
- Shafts (generally constituted of steels and in several sizes)
- Gears torque of several shapes, sizes, characteristics, properties
- Different commercial components (screws, nuts, bolts, bearings, and so on)
- One or more female adaptors that can be connected to different extensions.

The box is the external container of the reducer and the support for the internal components that form it. Shafts and gear torques are the basic components of the reducer and are responsible for the real transmission of motion and power in the ways envisaged by the plan for the specific reducer model. All the remaining components are for support, like for example bearings, stoppers, splines and so on.

Some of the above mentioned components can be produced inside COMER (responsible for product design test and assembly) and others are produced by partners of COMER, in particular both the VE partners usually produce boxes. COMER subsequently assembles, tests and delivers the final product to the customer.

With regards to companies working in the mechanical field the aspects connected with communication, information exchange and co-design are the most relevant. The main problems to be solved for the Supply chain are summarised below (the experiment focuses on these problems to validate the proposed ICT infrastructure):

- **Communication of data:** document and information exchange is very critical, especially during the product design phase: several documents have to be exchanged between COMER and its VE partners and their different formats can cause many problems and delays
- **Negotiation process:** negotiation between COMER and the partners became a critical issue. Especially for complex and new products it is fundamental to reduce delay, misunderstandings and errors
- **Quality aspects**

The principle results and benefits to be derived from the adoption of the BIDSAVER ICT platform are:

- *Speed of communication for orders and documents* to the partners
- *Improvement in the management of drawing releases*

Monitoring of orders and project progress (for the VE leader, COMER)

The overall description of the ICT platform designed and adapted to support the above-mentioned activities of the VE is given in the following paragraph. The Supply chain pilot case started in September 2001 and finished in June 2002.

3. BIDSAVER ICT PLATFORM DESCRIPTION

The ICT infrastructure for the support of the VE consists of three modules: BIIM, BBS and ICA. The modules have been properly integrated to provide a complete set of tools to manage the entire life cycle of the VE, from its constitution (partner information, contractual matters and so on) to its implementation (Design and Production phase).

It should also be envisaged that the ICT infrastructure obtained has to support both the SMEs and the Business Integrator activities, which are quite different. The SMEs forming a Virtual Enterprise need tools for the efficient exchange of documents and information (in different formats, especially during the design phase), while during the Production phase they need to manage orders, to track order status and to communicate the progress of the project. The Business Integrator role is quite different: the BI has to help the VE partners to find new partners for the production of specific components, to organise the VE processes and product structure and, especially during the Production phase, it has to monitor the project progress. As a result the VE partners and the BI should have different functions, views and interfaces at their disposal.

Another relevant point is that the SMEs forming the VE have their own existing legacy systems and the ICT infrastructure developed must be compatible with all of these and require a minimal effort of integration.

The functions provided by each module of the platform are described in the following paragraph with the implemented architecture of the system.

3.1 ICT Platform components

3.1.1 *BBS Module*

The BBS (Business Breakdown Structure management module) is designed primarily to support the VE constitution phase (see Figure 1), and contains all the necessary information to manage:

- The product, (with the Product Breakdown Structure, PBS) its changes and evolution (from the structure point of view).
- The processes involved in obtaining the product (WBS: Work Breakdown Structure) and possible changes there in. In particular, the defined VE process model can be stored and managed through the BBS.
- OBS (Organisational Breakdown Structure): the BBS stores and facilitates the management of partner information: basic and identification information (name, address, role in the VE and so on), performed phases within the VE processes.

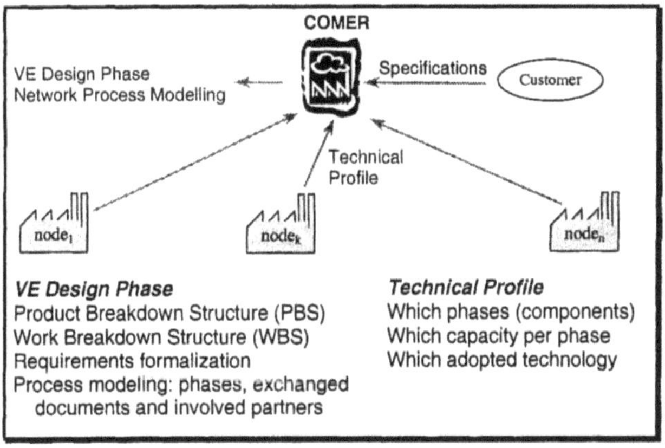

Figure 1 -VE Constitution

3.1.2 BIIM Module

The BIIM (Business Information Integration Module) is the main support tool for VE configuration (see Figure 2) and operation (VE Design phase and Production phase) with additional Performance evaluation tools (see the following Figures for a schematic representation). The BIIM allows the automatic Management of network activities.

VE configuration

Before starting the VE operation, the Virtual Enterprise needs to be configured in terms of Task distribution among the VE nodes and selection of the best partner for each operation. Once assigned the tasks, each partner can accept or refuse the assigned tasks, so a Negotiation phase and an alternative task distribution may be required.

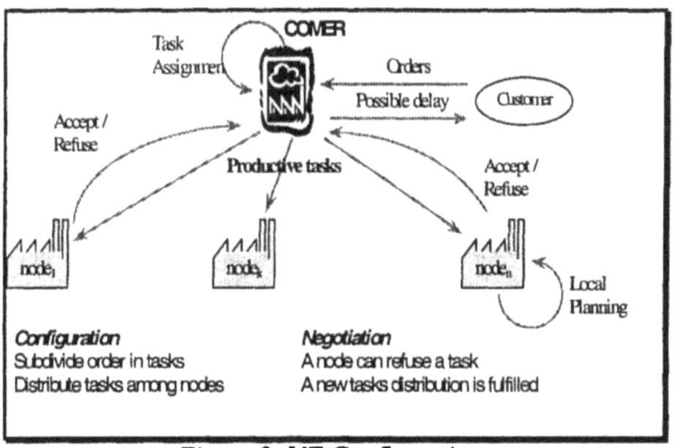

Figure 2 -VE Configuration

VE Operation

Once the task Assignment phase is successfully concluded, the real VE operation can start and each VE partner (or node) can start its tasks. During the VE operation the nodes have to periodically inform the network leader (COMER or the BI) of the task status and progress (see Figure 3). In particular, the problems encountered have to be immediately communicated to the network leader, to allow for prompt reaction in order to contain and minimise the impact on the remaining phases and on the delivery date of the final product. A mechanism for the modification and notification of tasks is required, to inform all the partners involved of any changes introduced into the work schedule.

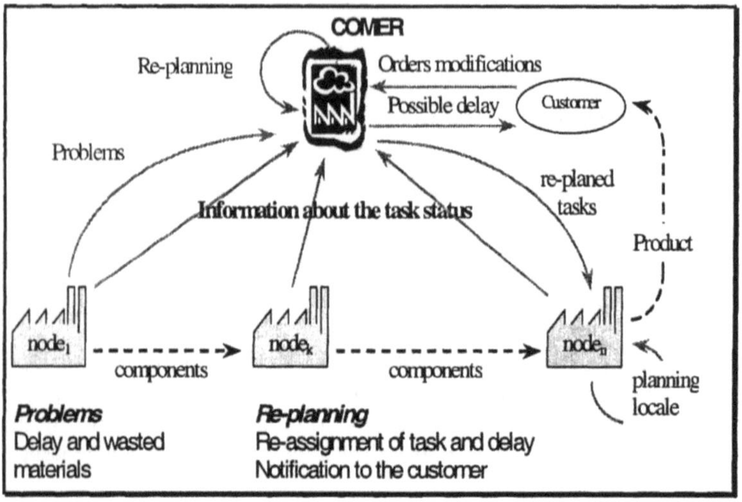

Figure 3 -VE Operation

3.1.3 ICA Module

The ICA module (Information Capturing Agent) supports the Search for partners through the Web. In particular, a robot (mobile agent) searches on the Internet for partners, according to pre-defined profiles by investigating web sites.

This module is mainly used by the Business Integrator while searching on the Internet for new potential partners based on the technical profiles provided by COMER (Supply chain leader), to cover the required roles and to be integrated in the Virtual Enterprise.

The resulting list of potential partners (with their relevant characteristics) is evaluated by COMER to select the best partner to be contacted and to start the negotiation phase. The search results can also be stored in a Data Base for further investigations.

3.2 BIDSAVER PLATFORM architecture

A wide ranging survey on the available commercial SW tools has been performed to find the best software on the market to fit the described requirements, and to be integrated to form the complete BIDSAVER ICT platform (see the list in Table1).

Table 1 – Commercial software on the market

Product	Company
Ematrix	MatrixOne, Inc.
Smart Team	Smart Solution Italia S.r.L.
Windchill	PTC
TeamPlay Concentric Suite	Primavera Systems, Inc.
Unigraphics	Unigraphics, Inc.
JetForm Corp	JetForm.
DSTInternational	Oper Product
Mindwrap Inc.	Optix
CAPS Inc.	Composite Administrator
Strategic Asset Management Solution	Socrates
Citadon	Citadon ProjectNet
Microsoft	Microsoft Project 2000

After an in depth analysis of the reported products UG iMAN demonstrated to be itself the one that best fits the user requirements (the missing Project management functions have been provided by integrating it with MS Project 2000). The iMAN solution has been selected and properly integrated also with the ICA module (which is a specially developed SW based on Web agent technology).

In particular, UG iMAN has been selected also because Unigraphics has a wide spread presence in Europe and several customer installations. In fact, UG solution provides also Unigraphics CAD software which in addition to being one of the most popular in Europe is also obviously integrated with the iMAN package (which is a PDM tool).

Customization for different networks can easily be achieved using the Application Programming Interface that iMAN provides through a software toolkit.

BIDSAVER can better support dynamic cooperation requiring a higher level of information and data structure complexity, especially in terms of configuration and product description features. The main targets of the BIDSAVER platform are large enterprises (in general the VE leader or a service centre) with complex products and processes to manage. In fact, the costs of iMAN licences are high (they start from 35.000 Euros per user) and the application needs a high performing server to run. The advantage is that all the VE partners will be able to use the system without installing any local program, simply by using the Browser and free of charge (see Figure 4).

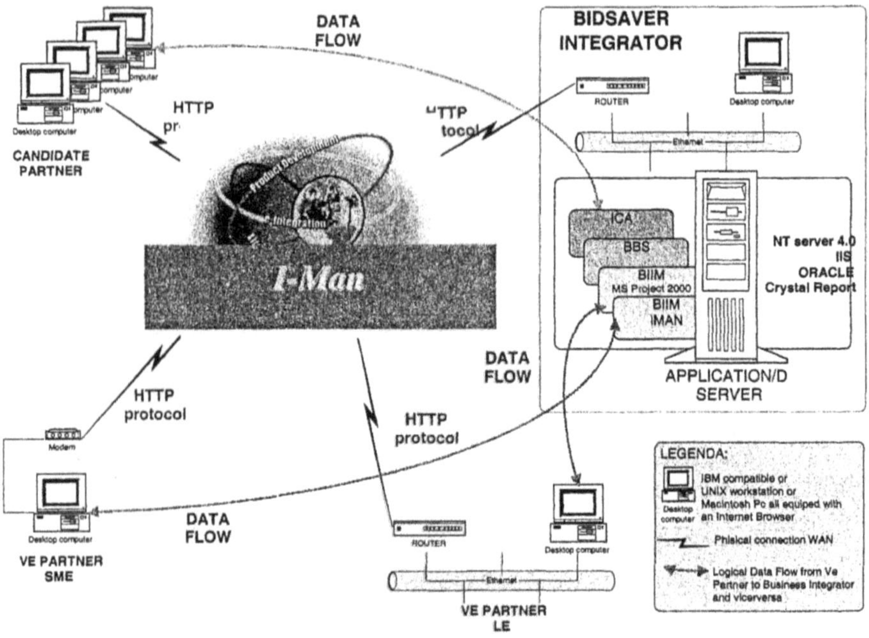

Figure 4 - BIDSAVER architecture

4. CONCLUSIONS

BIDSAVER delivered an ICT prototype for the Virtual Enterprise information system, leveraging on e-Commerce and Internet technology.

The system is constituted by a number of commercial applications and three main management modules:

- A **Business Breakdown Structure management module** (BBS), allowing the management and evolution of VE, both during the constitution phase and during the operation phase

- An **Information Capturing Agent** (ICA), performing the actual search for potential partners and providing updated information on co-operation opportunities during VE operation.

- The **Business information Integration Module** (BIIM), in charge of ensuring integrated management of functions needed for VE operation.

The prototype is based on pre-selected commercial solutions than can be configured according to VE requirements and improved through accommodating specific software functions necessary to the success of the VE operation.

The chosen SW tool is UG iMAN, which demonstrated to be very useful to support the WorkFlow Management functions (the missing Project management functions have been provided integrating it with MS Project 2000).

The iMAN solution has been properly integrated with the ICA module, a special SW developed by Eletel (Greece) based on Web agent technology. Customisations for different networks can easily be done, using the Application Programming Interface that iMAN provides trough a software toolkit.

The new ICT platform has been tested and validated on the filed by two real cases: The micro-satellite and the Mechanical Supply Chain Virtual Enterprise. The final evaluation of the results achieved during the Mechanical pilot case confirmed that the features and services of the BIDSAVER platform are unique with respect to other software packages specifically addressing cooperation among companies. These features can be summarised as follows:

- Management of complex product data structures;
- Dynamic management of project collaboration, in terms of product reconfiguration, process definition and partners' management.

The experiment has been carried out and organized with the main objective of testing, improving, tuning and validating the BIDSAVER SW platform. During the experiment the Mechanical VE members used the new SW platform for managing the co-operative and distributed activities related to the gearbox production (the product of the network chosen for conducting the test case).

The relevant processes defined and experimented are:

- **Purchase Requisition**: To allow COMER choosing the right supplier for the needed products.
- **Order Communication**: To send an Order to the chosen partner.
- **Exception management**: To manage possible problems met by a supplier while executing an order.
- **Non-Conformity Communication**: To allow COMER managing the lacks in Quality of the provided products.
- **General information exchange**: To allow the suppliers sending general information and documents.
- **COMER Exception**: To manage possible changes requested by COMER in an order execution.

COMER has been very satisfied of the advantages coming from the data sharing: thanks to iMAN, the data base containing all the processes and information, is shared with the other partners of the network and the information is available to all the authorised people everywhere and in any time. Furthermore, the introduction of iMAN significantly improved also the quality of work of the internal departments of COMER (especially the Design department). The new system increased also the traceability of information, reducing drastically the waste of time and costs for sending drawings in paper format (see Table).

Table 2 – Quantitative parameters at the conclusion of the pilot case

Parameter	Final Value	Notes
Data Sharing	EXCELLENT	When a new document is approved, each partner can read the information in real time, without delays and without possibility of mistakes, since they are

		recorded in the database.
Communication Traceability	EXCELLENT	Thanks to the Bidsaver ICT platform, each process is recorded, as well as each decision taken by the different partners and all the relevant comments.
Improvement Of The Internal Organisation	EXCELLENT	The platform improves the quality of work both from external and internal points of view: the documents are saved only once and are shared; so all the different partners and users read the same information.
Time Saved For Sending The Same Data To Different Partners	Up to 80%	It is no more necessary to print documents several times for sending them to different partners: they all are available in iMAN and to send them again to another partner takes only few seconds.
Efficiency In The Management Of The Quality Documents	EXCELLENT	The traceability has been improved and all the processes are recorded.

5. REFERENCES

1. Hirsch B. E. et al. Decentralized and collaborative production management in distributed manufacturing environments. Sharing CIME Solutions, Ios Press, 1995.
2. Filos E. and Vaggelis K. Ouzounis. Virtual Organisations Technologies, Trends, Standards and the Contribution of the European RTD Programmes, Berlin, 2000.
3. Bonfatti F., P.D. Monari, P. Paganelli. Co-ordination Functions in a SME Networks. BASYS '96 Intern. Conf. "Balanced Automation Systems II", Lisbon, 1996
4. IST-1999-10768 Bidsaver Annex I, Description of work, 1999
5. BIDSAVER Deliverable D1: Pilot cases scenario and Virtual Enterprises General requirements, 2000
6. BIDSAVER Deliverable D8: ICT architecture specifications, 2000

MANAGING VENDOR SUPPLIED MODELS OF PRODUCTION FACILITIES

Kevin Caskey[1], Kamel Rouibah[2], Henk-Jan Pels[3]
[1]School of Business Administration
State University of New York at New Paltz
caskeyk@newpaltx.edu
[2]College of Administration Science
University of Kuwait
krouibah@cba.edu.kw
[3]Faculteit Technologie Management
Technische Universiteit Eindhoven
H.J.Pels@tm.tue.nl

Modern production facilities are created by integrating subsystems supplied by several manufacturers, using many different models when creating their subsystems.

Our research goal is to integrate and manage these diverse sub-models in order to design and improve the facility itself. These applications include Product Data Management (PDM), CAD, models of robot dynamics, material flow, and operations planning.

This paper focuses on managing the many models and versions of models used throughout the life-cycle of the facility. In a task driven environment, an integrated model must be built with the sub-models and versions appropriate to the task.

Our work reflects the needs and constraints of the partners[i] in a multi-company distributed design environment, including an automobile firm, a robot manufacturer, systems integrators, and software vendors.

Keywords: Product Data Management, Model Integration, Distributed Design, Vendor Integration

1. INTRODUCTION

Modern production facilities are created by using sub-systems and services provided by many different independent vendors. The sub-systems are often designed with the help of different models. These models can describe the physical movement of machines such as robots, characteristics of the intended product, or descriptions of the intended manufacturing environment, such as production plans and schedules. Once a facility is in use, there are also several different models describing the performance and interactions of the components, the product being produced, and the operational conditions.

Often, both during design and during operation, there are engineering issues that span the domain of more than one discipline. For example, to judge whether a proposed welding robot can handle the intended load would require information from robot design as well as characteristics of the components to be welded and of the planned assembly load through the factory. The robotics engineer will have a dynamics model of robot arm movement. The manufacturer will have several different models describing aspects of the intended product (CAD, PDM, etc.). The manufacturing engineers will have models of the expected flow of materials and of the intended schedule of operations. Besides representing different engineering domains, this information will often come from different companies. In the above example these would be the robot manufacturer, the turnkey supplier, and the end-user.

The following sections will present the need for integrated models, dimensions of the model development, our approach to model management, and adapting a PDM system to implement this approach.

2. NEED FOR INTEGRATED MODELS

A modern production facility is created using the components and skills of many suppliers. These companies co-operate to deliver components and ultimately the production facility itself (see Figure 1). An example of a production facility may be an automobile assembly line. Within this facility, a subsystem could be a spot-welding cell, and components within the cell could be welding robots.

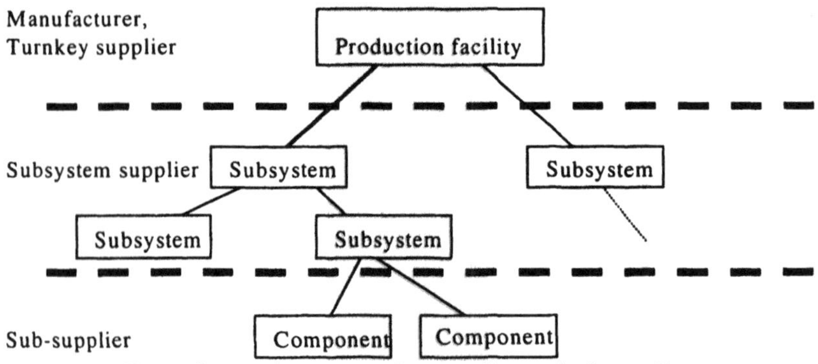

Figure 1: manufacturer, turnkey supplier, and sub-suppliers

The vendors of components often use models and simulations to test, and fine-tune their components or sub-systems before they are built and then sent to be integrated into the next higher level. It is cheaper and quicker to test and improve models of these sub-systems in a virtual world. However, once the sub-systems are sent to the highest level, final integration is done in the real world. Just as it is attractive to test and improve the components first in a modelled virtual world, it would also be so for the final integrated product.

Testing the entire production facility in a virtual world requires a model or simulation of the entire facility. It would be quite a task to create a complete model from scratch, though this is sometimes done. Much modelling effort may be spared if it were possible to combine the models already created by the vendors for their performance analyses (e.g. Klußmann & Caskey 1998).

However, building integrated models in an environment where the sub-models evolve requires an extensive effort in model management. The next section will describe how these models relate to their environment.

3. DIMENSIONS OF MODEL DEVELOPMENT

Models developed as part of factory planning reflect several aspects of the planning environment. Among these aspects are: Project phase, Project work domain, Document status, Model status, and Facility status. We will now discuss these aspects.

3.1 Project phase

A major engineering effort, such as the design of a new factory, evolves through defined project phases, such as the seven (0 through 6) in Figure 2:

0- Preliminary planning phase
1- Offer phase
2- Engineering phase
3- Design phase
4- Realization phase
5- Start-up
6- Running/changes phase
7- Documentation/end of project phase

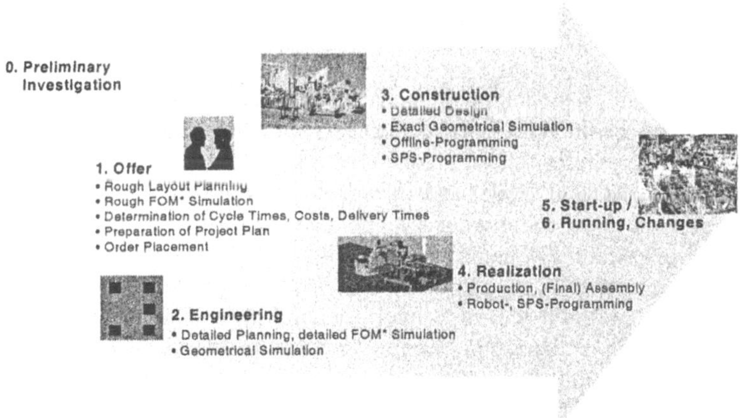

Figure 2. Phase model for plan development.

3.2 Project work domain

In traditional sequential engineering each project phase results in a complete, consistent and released specification or model of the target facility, each phase taking a particular view. Early phases would contain a high level of abstraction, later phases a deeper level of detail. For example, functional design, focussing on the functional aspect, occurred before detail design, that focuses on the geometrical aspect. In a concurrent engineering environment, the evolution is no longer strictly serial. These work domains overlap to allow the benefits of concurrent engineering, including better product functionality, manufacturability, and a quicker time into service. In order to keep the project manageable and to allow for concurrent engineering, the several complete, consistent models have to be separated from the project phases. Therefore the term engineering domain [Eerens, 1996] is a suitable term. At the end of the project each domain must hold a complete and consistent model of a specific aspect of the facility. Domains are ordered in the sense that each following domain must implement the elements of the previous domain [Helms, 2000]. The choice of domains is merely a matter of 'separation of concerns'.

There is no longer a one-to-one relationship between work domain and phase.

3.3 Document status

Documents in a complex design process go through a number of life cycle steps in which their contents evolve from incomplete and uncertain to complete and definitive. The status of a document indicates the current step or the last step that has been completed with success. Thus the status of a document (for example: identified, in-work, in review, released) reflects the acceptance of the set of engineering decisions embodied in the document. When changes are made to a document, a new version is created with at the lowest status level. The previous version keeps its current version, so status is version dependent.

The computer files that contain the software code of models are considered to be documents also.

3.4 Model Status

A model is a virtual presentation of a part of the facility. The model must be distinguished from the document that contains its specification or the software code relating to it. The reason for this is that the model goes through a different life cycle than the document, so that model and document can have different versions with different statuses. Typical statuses for a model can be: identified, in work, running, validated, released. Models of facility parts can be aggregated into integrated models of lager systems, such as cells or the whole facility. The selection of the right versions of sub-models with the right interfaces and input data sets, depending on the status and version of the facility and the purpose of the simulation, is the core task of model management.

3.5 Facility Status

Just as the project evolves through phases, the facility itself will evolve. The knowledge gained in facility operation, and the changing management goals at different stages in the life-cycle of a facility will impact the use of models and the building of integrated or composite models.

3.6 Product status

The designers of the production facility need up to date knowledge about the product to be processed on the facility. Therefore the products and their components must be represented in the model management facility. Products have their own versions with specific statuses.

4. MODEL MANAGEMENT

The above sections introduced dimensions under which models can be classified. We will now describe our approach to model management. The applied nature of the consortium imposes criteria around which our approach is formed: use existing applications, develop the models locally and then run them centrally, and use models at a maturity appropriate to the phase of engineering design. These will now be described in a bit more detail. We will also present how our research methodology fits into this user-driven scheme.

4.1 Task within Phases

Engineering tasks drive the use of integrated models. Each project phase comprises several activities. In addition each activity has inputs and outputs. Table 1 contains examples of such activities. The problem addressed by Model Management is that the models used by the task evolve during the project, so that different versions of each model exist. The task of Model Management is to select the right combination of models (with the appropriate version and status) and input data sets for each specific task. This is not a trivial problem since the right version is not necessarily the last one.

4.2 Conditions imposed by Industrial partners

The work described here is driven by the needs of a consortium, especially the end-users. The needs dictate which models must interact under what circumstances. The end-users intent and their operating environment also impose conditions on the approach taken. For example, the sub-models will be developed using the software tools the firms currently use, and will be developed by the party responsible for the component and view (see Figure 1).

The solution provided should not require end-users to change the products they use for their major applications. For example, the automobile manufacturer uses CATIA for its CAD system. Any solution that would require them to switch to another system would not be acceptable. By extension, the system vendors are not partners in the project, so changes to the packages, such as CATIA, are also not acceptable.

Table 1. Examples of activities, roles, input/output and resources per phase

	Activity	Input	Output	Role	Technical resources
1	Assembly sequence planning	Product data version 1, Factory layout, Tech. info.	Manufacturing bill of material	Process planner, (Assembly planner)	Digital mock-up tool (DMU), Production planner (PP), CAD
2	Detailed planning	Product data version 2, Factory layout, Technical information (e.g. process data, material)	Manufacturing bill of material, Assembly sequence	Project engineer, Project planner, Process planner, Material flow simulation expert, Logistics expert, Tech. expert, Assembly planner, Factory planner	DMU, PP, Office productivity software, Manufacturing process planner, 2D/3D-layout sys.
3	Detailed design	Product data version 4 (3D geometry), Factory layout, BOM, Assembly sequence	Optimized line and factory layout, Necessary standard machine tools	Resource design engineer, Factory planner	CAD, Manufacturing process planner, PDM, ERP
4	Online program ming	Existing code, Cycle time, Assembly sequence, Sequence of optimization, Process in/out, data in/out	Software validation, Verification	Process planner, Work cell simulation expert	Human-machine interaction model, Off-line programmer for PLC
5	System improve ment	Product data version 5, Factory layout, Cycle time, Throughput time, Buffer size	Modified: cycle time, -throughput time, - (buffer) capacity, - resources	Project engineer, Process engineer, Process planner, Tech. expert, Resource design engineer	CAD, PDM, Human-machine interaction model,, Work cell modelling tool
6	Detailed design	(New) product data version 7, Factory layout, (New) BOM, (New) assembly sequence	Optimized line and factory layout, Necessary standard machine tools	Resource design engineer, Factory planner	CAD, Manufacturing process planner
7	Final drawings design	Factory/Line Layout, Details of tools and machinery, Collection of all drawings from all suppliers	Complete set of drawings, Simulation models according to OEM specification	Resource/product design engineer	CAD, Work cell modelling tool Robot instruction program

4.3 Use appropriate sub-models

As the design project moves through the phases the maturity of the several models evolve. For example, a material flow simulation may be rather rough in early design phases and become quite exact and detailed just before the design is frozen. While the rough simulation would be too crude to be used to support engineering decision-making in the later phases, it is sufficient to support earlier analysis. Whatever approach we propose must be able to provide models adequately mature for the tasks in a given phase. For many decisions it would not be realistic to wait until all models are in their most mature form.

5. EXPANDING PDM DOCUMENT CONTROL

We are building our model management system upon a commercial PDM system. An overview of functions supplies by commercial PDM systems can be found in Helms (1999). While this provides a start, it does not provide a complete solution to the multi-dimensional view defined in section 3. The document management function of the PDM system enables us to store models as versioned documents and to control access, so that non-authorised parties cannot access models. The product structure management function of the PDM system enables us to store and maintain the hierarchical facility and model structures that are needed to select the right components for integrated models. The workflow function allows us to define the different model statuses and to link them to project tasks.

Managing versions of software components is not a new topic. However, two issues must be addressed in this setting. The end users require the models to be developed locally and then sent to the other participants. Engineering analysis involving several sub-models will then be performed locally. Development work on sub-models could lead or lag design phases. In the above example, this could impact in two directions. Say the party performing detailed *planning* is indeed working in Phase 2 but has only received version 1 of the product data. Will the analysis be valid? On the other hand, what if the product data is already in version 2 while the Phase 1 assembly sequence is still to be done. We suspect that the first case would indeed cause problems but the second case not. It is often believed that a more mature version of a sub-model can always be substituted for an earlier version. The reverse should not be true because some information would be missing.

However, it can also be that a newer version corrects an error found in an earlier version. In this case, knowledge of the analysis task would be required to determine which versions, if any, can be assembled into a composite model.

In the integration phase of the project, we will link the project phases to document release and the tasks to appropriate assemblies of sub-model versions. This phase will also tell us whether the upward compatibility is indeed generally true and whether it is possible but raises difficulties and costs.

6. CONCLUSIONS

Due to the evolving character of models during the design and installation of complex production facilities, the management of consistent integrated sets of models is a non-trivial task. This task is called Model Management. Commercial PDM systems provide the basis of functionality to support the model management task. Further research is directed towards the development of a detailed data structure model of the model management problem and specification of procedures to derive consistent sets of models and input for specific tasks in the project.

7. ACKNOWLEDGEMENT

Much of the development described in this paper is being performed within the VIDOP (Vendor Integrated Decentralised Optimisation of Production Facilities) project. This is a shared cost project partially supported through the European Commission's Competitive and Sustainable Growth programme. The consortium partners are:

Germany:	KUKA Schweissanlagen GmbH, DaimlerChrysler AG, Universität Karlsruhe
Spain:	INGEMAT,S.A., Fundacion ROBOTIKER
Portugal	EFACEC AUTOMACAO E ROBOTICA, S. A., Faculdade de Engenharia da Universidade do Porto
Italy	UBK Italia Srl
Israel	Tecnomatix Technologies Ltd.
The Netherlands	Turnkiek Technical Systems, Technology Management Contract Centre, B. V

The partners acknowledge the participation of the European Commission.

[1] The partners are co-operating within the VIDOP research consortium (see Acknowledgement)

8. REFERENCES

Erens, F.J., The Synthesis of variety – developing product families, dissertation Technische Universiteit Eindhoven, 1996, ISBN 90-386-0195-6

Helms R.W., The PDM Function Model, deliverable 5.1 RapidPDM project (Esprit 26892), September 1999.

Helms, R.W. , Pels, H.J., Elzen, R. v.d., Simulation Product Development Processes from a PDM point of view, proc. TMCE 2000 Conference, Delft, April 2000.

Klußmann, J, and K. Caskey, Increasing the usability of material flow simulation models through model data exchange, *Simulation Practice and Theory*, 6 (1998), 35-46.

Norman, M., Tinsley, D., Barksdale, J., Wierholm, O., Campbell, P. and MacNair, E., Process and Material Handling Models Integration, *Proceedings of the 1999 Winter Simulation Conference,* 1999.

PART FIVE
SUPPORTING TECHNOLOGIES

ESSENTIAL SERVICES FOR P2P E-MARKETPLACES

Diogo Ferreira[1], J. J. Pinto Ferreira[1,2]
[1]*INESC Porto,Campus da FEUP, Rua Dr. Roberto Frias 378, 4200-465 Porto, Portugal*
[2]*Faculty of Engineering U. P., Rua Dr. Roberto Frias s/n, 4200-465 Porto, Portugal*
dmf@inescporto.pt,jjpf@fe.up.pt

Existing e-marketplaces, built on traditional client-server architectures, severely restrict the scope and dynamics of B2B interactions. Peer-to-peer (P2P) architectures will provide far more decentralised infrastructures, while allowing a much wider range of business patterns to take place. The purpose of this paper is twofold: (1) to argue that interaction over a P2P network better resembles the way enterprises perform business with each other, and (2) to point out a set of essential services required to make P2P infrastructures fit for B2B exchanges. An overview of those services is presented, based on one of the leading P2P platforms (JXTA).

Keywords: *E-marketplaces, B2B e-commerce, peer-to-peer networking, JXTA*

1. INTRODUCTION

The rise of a global and ubiquitous infrastructure such as the Internet has brought speed and connectivity to the whole spectrum of B2B exchanges, from simple document delivery to complete and automated trading procedures. E-business possibilities have led to new business-to-business (B2B) models, particularly the *e-marketplace*. However, traditional Web-based, client-server architectures exhibit several limitations mainly because they are a centralised solution for a problem which clearly requires a distributed infrastructure.

On the other hand, the impact and success of completely decentralised, peer-to-peer (P2P) solutions has placed this paradigm as a better alternative to cope with B2B trading requirements. Recently, several P2P infrastructures have been proposed (Gnutella, FreeNet, JXTA, Jabber, etc.). Still, these infrastructures lack essential services allowing companies to use them as business platforms. This text proposes and describes the implementation of those services, which will support B2B exchanges over P2P-based e-marketplaces.

2. THE TRADITIONAL E-MARKETPLACE

A significant part of recent B2B developments has been devoted to *e-marketplaces*, Web portals gathering both buyers and sellers and fostering business transactions among them. Though there are e-marketplaces of all sizes, the typical ones focus on a particular type of business (for example automotive parts, financial funds or chemicals). They are also *neutral*, i.e., they are set up and run by an independent third party. E-marketplaces usually comprise two main areas: one dedicated to the member suppliers advertising selling offers and another to member buyers where they advertise purchase needs. Additionally, there is usually support for auctions and reverse auctions, with buyers bidding for supplier offers or sellers attempting to fulfil demand at the lowest price, respectively.

Typically, a prospective seller publishes offers while a prospective buyer searches for products or services by looking up offers on the marketplace database. If an interesting offer is found, the entity running the e-marketplace will work as a *broker*, mediating contact between buyer and supplier. Both buyer and supplier will always interact solely through the e-marketplace, and never directly. In the end, the broker will usually charge a service fee taking the form of a transaction value percentage.

Thus, the traditional e-marketplace exhibits the following characteristics:

Centralised repository. An e-marketplace is a centralised information repository where buyers and sellers publish and subscribe information of their interest.

Opaque interfacing. Though buyers can browse offers, the e-marketplace will not reveal trading partners' identity in order to assure its own intermediary role.

Static information. The information on that repository is typically static in nature, declaring what a seller is able to provide for what price, anytime. Because an e-marketplace is opaque, offers cannot be tailored to specific customers.

Bottlenecked structure. The e-marketplace is the single interface for all trading partners, and therefore it must be prepared to handle the simultaneous transactions of all registered users.

Loosely connected. Being a single interface for all business transactions, it is also a potential point of failure. In the case where an e-marketplace ceases operation, its trading community will become unavoidably scattered.

Fixed exchanges. All B2B exchanges obey to a set of predefined rules of what a business partners should provide and expect. These rules may be based on B2B frameworks such as OBI, RosettaNet, or cXML (Shim et al, 2000). Because these rules are imposed by the e-marketplace, it may be quite different from what would happen if partners were to trade directly.

Price-based decisions. Buyers have to make decisions mainly on price. But choosing the supplier with the lowest unit cost may not always yield the best results.

Non-iterated agreements. Because B2B exchanges opaque, fixed and price-based, there is little support for agreements attained by iterated phases of negotiation with proposals and counter-proposals.

Short-termed partnerships. Because the e-marketplace is opaque, it is not possible to establish long-term business partnerships or initiatives.

Service charges. The e-marketplace lives on the premise that it is able to bring together buyers and suppliers, and attain successful business transactions. But imposing a fee based on transaction value may be dissuading for prospective trading partners.

3. DECENTRALIZING B2B EXCHANGES

The services provided by the traditional e-marketplace may be approached from a P2P perspective as well, bringing numerous advantages. On a P2P e-marketplace peer nodes represent both buyers and sellers. Peers may search, connect, and exchange data with other peers. In the P2P network, sellers may publish, advertise, or by other means provide information about their offers. Sellers rely on the P2P infrastructure to convey that information to other peers. Buyers, on the other hand, rely on search capabilities to locate and retrieve offers, and on the network infrastructure itself to connect and interact with sellers.

Up to now, interaction over a P2P network has been usually confined to two phases: (1) the search-and-find phase when one peer contacts several others in order to locate desired information, and (2) the connect-and-retrieve phase when the same peer obtains the intended data from another peer. For the purpose of P2P e-marketplaces this is still limited, because peers must be able to interact in a bi-directional way: first while setting trading conditions, and afterwards when attaining the previously agreed exchanges. Notwithstanding, P2P networking provides one-to-many and (possibly secure) one-to-one communication which, associated with powerful search and transfer capabilities, can significantly improve the development of business relationships on e-marketplaces.

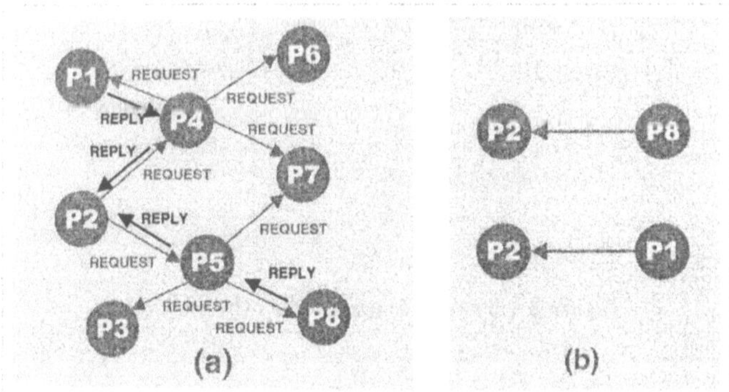

Figure 1 – Search (a) and retrieval (b) over a P2P network

This same behaviour opens a wide range of possibilities for more complex business patterns, such as the ones specified by B2B frameworks (once again like OBI, RosettaNet, cXML and others). Moreover, it allows peers to establish contracts directly, and P2P interaction is only limited by the reach of services that peers are able to deploy on the network. There may be a service for performing transactions according to OBI, and another service to deal with procurement according to cXML. Still, and before that can be done, supporting search and negotiation services must be in place. Equipped with those services, a P2P infrastructure will allow companies to perform flexible and dynamic B2B exchanges on a decentralised, broker-free marketplace.

Without further ado, it can be argued that P2P infrastructures, together with a set of supporting services, can surpass all of the shortcomings of traditional e-marketplaces pointed out in the last section.

4. DISSECTING THE B2B TRADING LIFE-CYCLE

In order to identify the essential services that P2P e-marketplaces should provide one must delve into the individual phases of the B2B trading life cycle. Figure 2 presents those phases, illustrating two distinct business processes which, according to the CIMOSA architecture (AMICE, 1993), may be referred to as denoting the *system life cycle* and the *product life cycle*, the latter being nested in the operational phase of the former one. The product life cycle leads to B2B trading needs. For example, some production steps are to be subcontracted, or a supplier for a sub-component must be contacted, or a funding partner must be found. The system life cycle, which encompasses the whole B2B trading cycle, introduces requirements for the underlying B2B integration infrastructure.

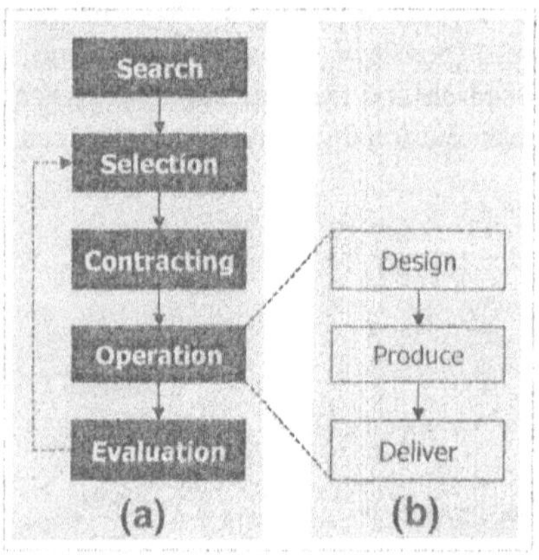

Figure 2 – System (a) and product (b) life cycles

4.1. Search

As illustrated in figure 2, the first phase is partner search. Here, companies use requirements drawn from the product life cycle to search for other key players. The search begins by issuing a search request. Specification of a search request may follow a format such as the *technology request* (TR) used to search for partners on the European Innovation Relay Centre Network (CORDIS, 2000). This format combines both human-readable information and standard business codes that allow automatic indexing of TRs. The counterpart is the *technology offer* (TO), a document following a similar structure but describing a product or service offer. For each TR, some or none matching TOs may be found. Both TRs and TOs are anonymous, specifying a contact point but concealing the identity of the company providing the document.

4.2. Selection

In response to a (one-to-many) search request, a peer will probably receive a set of matching offers, the point when the second phase – selection – begins. The selection phase is comprised of at least two smaller steps. The first step is pre-selection, i.e., immediately ruling out offers that seem unrelated or inappropriate. The remainder will be a set of candidates from which a choice cannot be made without further information. The second step is therefore to contact the peers referred to in the remaining TOs and, disclosing the identity of buyer and seller, proceed with further inquiries, on a one-to-one basis. Iterated steps of contact and inquiry – negotiation – may follow until the candidate offering the best conditions is found. These conditions may be based on many factors other than price alone.

4.3. Contracting

Once the business partner is chosen, a legal contract must be established between peers. The contract, called a *trading partner agreement* (TPA), formalises the legal conditions under which the trade will take place. Typical TPAs contain mainly human-readable information stating the rights and duties of each trading party (Taylor, 2001). Recent efforts in devising appropriate TPA formats for B2B exchanges (Dan et al, 2001), (ebXML, 2001) have led to specifications with a strong technological bias, but irrelevant for legal purposes. Research projects such as ECLIP (Cavanillas and Nadal, 1999), eLEGAL (Hassan et al, 2001), and Octanc24 (ComNetMedia, 2001) have taken a step towards legal frameworks, but there is still little agreement as to how TPAs should be represented. This is more of a legal issue than a technological one, since the enabling technologies – languages (XML), protocols (S/MIME, HTTPS, TLS) and electronic signatures (digital certificates, X.509) – are all available.

4.4. Operation

The operation phase concerns the assignment of resources to operational business processes, as well as enacting and controlling the entire product life cycle. Some resources are internal to the company – such as employees, machines, and applications – while others are external – precisely the ones having been negotiated in the previous phases. During process execution interaction with resources should be done transparently, whether they are internal or external.

Therefore, the operation phase requires B2B front-ends to be properly integrated with the back-office systems of each company, i.e., real-time back-office integration is required, as opposed to batch-oriented integration approaches (Lamond and Edelheit, 1999). In practice, secure communication channels should be in place, as well as application wrappers or gateways allowing data to be exchanged between endpoint systems across the marketplace.

4.5. Evaluation

The final phase of the system life cycle is an evaluation phase when overall performance and the performance of each contracted partner are registered. These measures, together with a log of all activities will help the selection of future partners in forthcoming trading opportunities. Evaluation, which is a strictly internal process, is also a chance for improving both the operational process and the B2B integration services, which are the focus of the next section.

5. IMPLEMENTING P2P SUPPORT SERVICES

Several services are required in order to support the life cycle just described. First of all, a search service must match technology offers to technology requests. Then, there must be direct exchange mechanisms for peers to exchange TPAs and other documents. Additional services providing support for particular B2B frameworks could also be in place. Finally, B2B interactions must be integrated with each partner's internal business processes. This requires real-time back-office integration between the enterprise system and the P2P network, as well as the deployment of P2P-aware client applications.

While several P2P platforms exist, the JXTA project (Sun, 2001) seems to be the most comprehensive and flexible framework. In a JXTA network, peers communicate through pipes, a protocol-independent abstraction. Pipes may use TCP/IP sockets, IP multicasting or HTTP to establish a connection between two endpoints on the network, no matter how far apart. The connection between endpoints may not be direct and is usually attained through a sequence of JXTA *routers* and *rendezvous* peers that allow traffic to circumvent firewalls.

The powerful feature of JXTA is that peers may create and join peer groups where special-purpose services can be deployed. A service can be any functionality that implements behaviour on the P2P network. Services may be used for searching, interacting, or any other purpose that peer groups may find useful.

The JXTA infrastructure is based on a particular type of XML document called *advertisement.* All JXTA resources (peers, peer groups, pipes and services) are represented by an advertisement. JXTA already provides community services – particularly the Discovery Service – that allows peers to find those resources. For example, peers search for group advertisements, then join the group and search for service advertisements, which in turn may contain pipe advertisements allowing the peer to interact with the service.

Hence, the JXTA platform comprises an architecture that is quite appropriate for the development of P2P e-marketplace services.

5.1 Search services

One of the cornerstones of a P2P infrastructure is the ability to perform distributed searches on available content. It is thus not surprising that JXTA already provides its own search service – JXTA Search (Waterhouse et al, 2002). The JXTA Search service assumes there are three kinds of behaviour a peer may exhibit: information provider, information consumer, and search hub.

Information consumer applications send requests to the nearest search hub, which decides which providers to forward the request to. The search hub also receives replies and sends them back to the consumers. In many applications, a peer will act both as provider and consumer.

Peers will probably connect to different search hubs which, in turn, are connected across the JXTA network. Using JXTA Search requires, therefore, a certain number of search hubs to be running, and requires providers to register themselves on these hubs. Moreover, due to its emphasis on Web content, JXTA Search is built upon the Tomcat servlet container. Also, as of this writing, the search service requires an older JXTA release.

An alternative, lighter solution would be to build a search service based exclusively on the mechanisms provided by the JXTA infrastructure. In fact, and using the concept of peer groups and propagate pipes, it is possible for peers to submit search requests to other peers in their group, without intermediary hubs. Because the purpose of this search service is to find trading partners, it is called *Trading Partner Search Service* (TPSS).

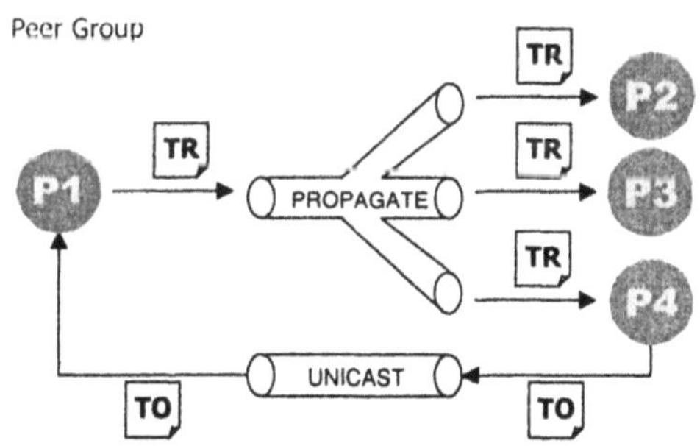

Figure 3 – The Trading Partner Search Service (TPSS)

The service is available under a group called "marketplace", with password-protected membership. When joining the group, a peer looks for a propagate-pipe advertisement that provides a one-to-all communication channel, and submits technology requests (TRs) using this pipe. At the opposite endpoints, the service matches the TR against the local technology offers (TOs) and returns the matching results. Each TR includes a unicast pipe advertisement specifying the input pipe through which the requester receives the matching TOs.

5.2 Negotiation services

Since TOs and TRs are anonymous documents, peers must have a way to exchange additional information, should a particular TO turn out to be interesting. That information is conveyed through a *Trading Partner Information Service* (TPIS). The TPIS works through a secure unicast pipe, on a request-reply basis. Each TO the search service (TPSS) returns includes a pipe advertisement for accessing the TPIS on the corresponding peer. Through the TPIS, peers exchange information such as company profile, available infrastructure, and preferred partners. Because the TPIS is a generic information exchange service there is no mandatory format for the information exchanged. Its purpose is to serve as a decision-support service during the selection phase.

Every message exchanged through TPIS should, however, include a pipe advertisement, this time for interacting with a contracting-support service, the *Trading Partner Agreement Service* (TPAS). This way, and after the selection phase is concluded, the chosen partner can be contacted in order to establish a TPA. The TPAS requires a secure unicast pipe just like the TPIS but, in contrast with the request-reply behaviour of TPIS, the TPAS allows iterated exchanges until a (digitally signed) TPA has been achieved.

5.3 Operation services

A TPA includes the pipe advertisements needed for the operation phase. Through these pipes, local applications at each peer can communicate directly, sending from or bringing messages to the local information system. This is achieved through the *Trading Partner Exchange Service* (TPES) that supports exchanges according to one or more B2B frameworks. A similar service interface is provided by the *Trading Partner Task Service* (TPTS) allowing interaction with the local resources that are accommodated in the P2P infrastructure as well.

6. CONCLUSION

Available P2P platforms are still not mature enough to support B2B exchanges such as the ones that take place on e-marketplaces. But once the proper services are in place, P2P architectures will probably surpass every other architecture by providing an infrastructure that better resembles the distributed nature of B2B trading.

This paper has pointed out some services for that infrastructure, that current P2P platforms still lack, assuming a particular B2B exchange life cycle. Clearly, though, more developments are needed in this field in order to achieve a comprehensive B2B framework of its own and, ultimately, to show the feasibility and suitability of the P2P approach.

Acknowledgment. The first author highly appreciates the support of the Foundation for Science and Technology (http://www.fct.mct.pt), which is funding his research.

7. REFERENCES

1. AMICE, ESPRIT Consortium. CIMOSA: Open System Architecture for CIM. 2nd ed, Springer-Verlag, 1993
2. Cavanillas S, Nadal AM. Deliverable 2.1.7: Contract Law. http://www.eclip.org/documents/deliverable_2_1_7_contract_law.pdf, 1999
3. ComNetMedia AG. OCTANE e-contracting service concept. http://www.octane24.com/oct_home/oct_project/pro_background/chart/chart2, 2001
4. CORDIS. IRC Public Website. http://irc.cordis.lu/, 2000
5. Dan A, Dias DM, Kearney R, Lau TC, Nguyen TN, Parr FN, Sachs MW, Shaikh HH. Business-to-business integration with tpaML and a business-to-business protocol framework. IBM Systems Journal, vol.40, no.1, 2001
6. ebXML Trading-Partners Team. Collaboration-Protocol Profile and Agreement Specification. Version 1.0. http://www.ebxml.org/specs/ebCCP.pdf, 2001
7. Hassan TM, Carter C, Hannus M, Hyvärinen J. "eLEGAL: Defining a Framework for Legally Admissible Use of ICT in Virtual Enterprises". In Proceedings of the 7th International Conference on Concurrent Enterprising (ICE2001), Bremen, 2001
8. Lamond K, Edelheit JA. "Electronic commerce back-office integration". BT Technology Journal, vol.17, no.3, July 1999
9. Shim SSY, Pendyala VS, Sundaram M, Gao JZ. "Business-to-Business E-Commerce Frameworks". IEEE Computer, vol.33, no.10, October 2000
10. Sun Microsystems Inc. Project JXTA, http://www.jxta.org/, 2001
11. Taylor ML. Trading Partner Agreement Sample. http://www.mltweb.com/ec/tpa.htm, 2001
12. Waterhouse S, Doolin DM, Kan G, Faybishenko Y. "Distributed Search in P2P Networks". IEEE Internet Computing, vol.6, no.1, January/February 2002

PROPOSITION OF A REPOSITORY FOR ENTERPRISE MODELING : (EMC) ENTERPRISE MODELING COMPONENTS

A. Abdmouleh, M. Spadoni, F. Vernadat
LGIPM – ENIM/Université de METZ
Ile du Saulcy F-57045 Metz Cedex1, France
a.abdmouleh@enim.fr
spadoni@enim.fr
vernadat@enim.fr

The design meta-models for the CIMOSA reference architecture are presented. These meta-models are formalized with UML (Unified Modeling Language) and configured with Design patterns. Patterns allow the reuse of reference models and give them solid foundations. Indeed, the CIMOSA constructs are written using generic and trade patterns concerning production means, physical flows, information flows, resources, ... UML is the most practical language to represent these patterns, thanks to its diagrammatic notations, which cover static aspects (e.g. objects, classes, components) and dynamic aspects (e.g. collaboration, scenarios, activities) of a system.

Keywords: CIMOSA, Repository, Pattern, Framework Meta-model.

1. INTRODUCTION

The current variations of the economic context force enterprises to become more flexible and reactive to be able to anticipate and adapt to frequent changes. The management of change relies on a good knowledge of the enterprise and its environment, the installation of anticipation mechanisms and by a modular structure to get a strong capacity of flexibility and adaptability. This process of change can be controlled and improved only by efficient use of enterprise modeling. This discipline is applicable to production enterprises of manufactured goods (continuous or discrete production) as well as to those providing services. It should be noted that the modeled processes can concern administrative *(or workflow)*, technical or support procedures.

To address these problems, foundations for a technical **repository** are proposed to support creation, sharing and collaboration between the model components of **CIMOSA** *(CIM Open System Architecture)* (CIMOSA 1994; Vernadat, 1996). These components result from **Design patterns** (Gamma, 1995) which are formalized in the form of a specific and unified language for enterprise modeling, called **UEML** *(Unified Enterprise Modeling Language)* (Vernadat, 2001). The proposed *Enterprise Modeling Components* (**EMC**) repository helps to solve modeling, sharing and collaboration problems between business users of an enterprise, and takes part in a broader chain of customer-supplier relationships of networked enterprises on large supply chain (Vernadat, 1999).

In this article, the CIMOSA framework and the EMC Repository based on CIMOSA meta-models are presented. The meta-models are formalized using UML (Unified Modeling Language) (OMG, 2001; Booch, 1999; Eriksson, 2000) and object patterns.

2. CIMOSA FRAMEWORK

CIMOSA provides a consistent architectural framework for both enterprise modeling and enterprise integration as required by CIM environments. It has been developed by the AMICE Consortium as a series of ESPRIT Projects (AMICE, 1993).

This CIMOSA modeling framework, also known as the CIMOSA Cube (Figure 1), consists of two parts:

- A reference architecture, and
- A particular architecture.

The particular architecture is a set of models documenting the CIM environment of a specific business entity from requirements to implementation.

The reference architecture is used to assist business users in the process of building their own particular architecture as a set of models describing the various aspects of the enterprise at different modeling levels. It is split into two layers: a generic layer providing generic blocks (i.e. basic constructs of the modeling language, their types, and instantiation and aggregation rules), and a partial layer consisting of a library of classified, re-usable partial models for some industry sectors (i.e. partially instantiated models that can be customized to specific enterprise needs). The aim of the modeling framework is to provide semantic unification of concepts shared in the CIM environment.

The CIMOSA modeling framework is based on three orthogonal principles:

- The derivation principle, which advocates to model enterprises according to three successive modeling levels:
    - Requirements definition to express business needs as perceived by users;
    - Design specification to build a formal, conceptual, executable model of the enterprise system ;
    - Implementation description to document implementation details, installed resources, exception handling mechanisms, and taking into account system non-determinism.
- The instantiation principle based on three distinct layers:
    - A generic layer containing generic building blocks and building block types as the elements of the modeling language to express any model (partial or particular);
    - A partial layer containing libraries of partial models classified by industry sectors to be copied and used in particular models; and
    - A particular layer containing particular models, i.e. company specific models of parts of a given enterprise.
- The generation principle, which recommends to model manufacturing enterprises according to four basic but complementary viewpoints:

- The function view, which represents enterprise functionality and behavior (i.e. events, activities and processes) including temporal and exception handling aspects:
- The information view, which represents enterprise objects and their information elements;
- The resource view, which represents enterprise means, their capabilities and management aspects;
- The organization view, which represents organizational units, decision levels, authorities and responsibilities.

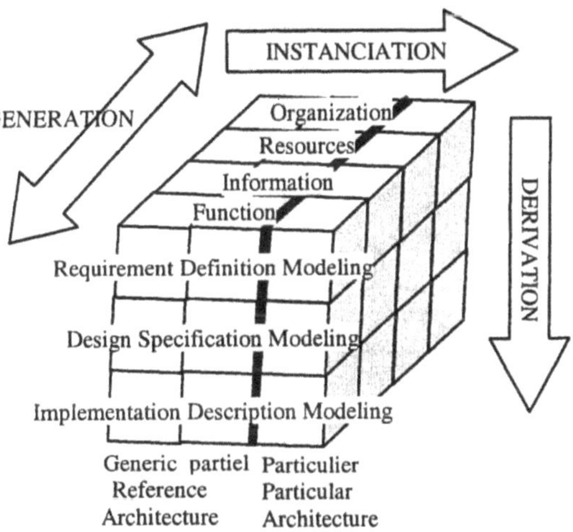

Figure 1- CIMOSA Cube

3. EMC REPOSITORY

3.1. INTRODUCTION

A repository is a warehouse for corporate data and information. Information is shared and reused by all users connected around this repository. In fact, each user can retrieve pieces of information he needs to manipulate them in his environment with various local existing applications. This means that a repository allows interoperability (Fairchaild, 1996) between existing applications used in enterprise. Repository information should be managed by an administrator, who is the only one empowered to modify, add and/or delete meta-information.

3.2. FUNCTIONALITIES

In the proposed EMC Repository, information items are implemented as components that describe meta-models and models. Meta-models can be shared by users (i.e. designers, personnel), according to access permissions, to model their particular models.

The repository components are modeled with UML, and by using patterns to improve their modeling with design knowledge and know-how.

In the next section, patterns are treated more deeply.

The EMC Repository allows management and sharing of components (i.e. generic building blocks and partial models) between CIMOSA generic and partial layers. The use case diagram (Figure 2) shows possible interactions between external actors and the repository as well as generic tasks, which can be achieved; By using a CIMOSA-based meta-model, a modeler can build particular models of an enterprise. These models can later be handled and shared on a day-to-day basis by various actors (e.g. modelers, designers, users, personnel and implementers) according to access permissions. These actors can create specific object models (e.g. manufacturing processes, design processes, management processes) useful for the work in their environment (e.g. simulation environment, design environment, ...) by using existing application systems (e.g. ARENA, ERP, ARIS Toolset, PDM, EDI, ...).

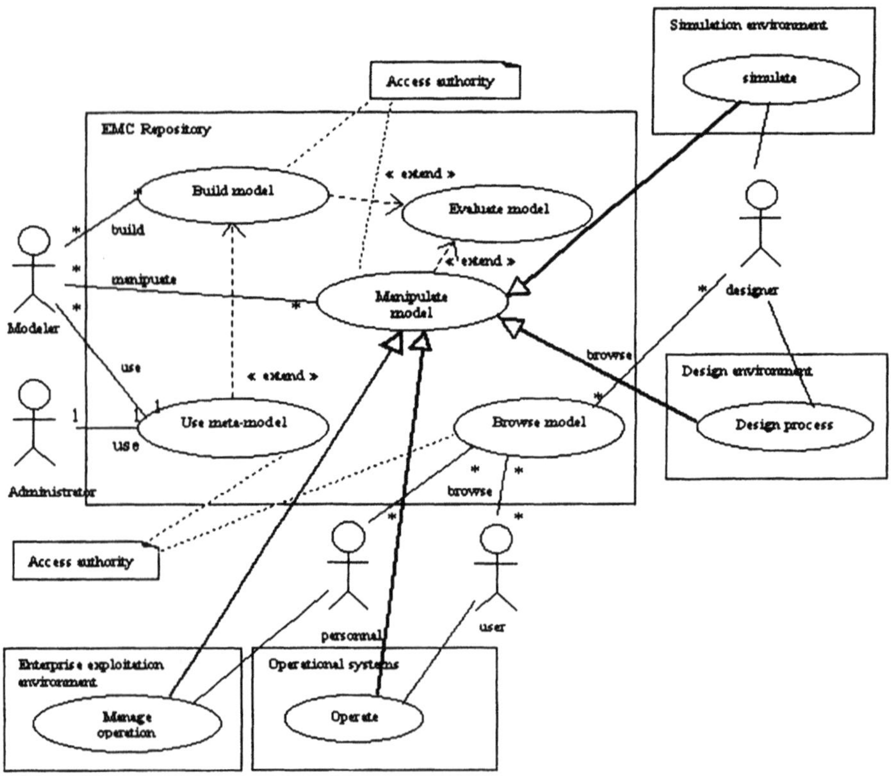

Figure 2 - EMC Repository use case

4. PATTERNS

A pattern is an abstraction of a domain solution, which can prove to be useful for other domains. It capitalizes a recurring problem of a domain and its solution, so as to facilitate reuse of this solution. Therefore, it constitutes a knowledge and know-how capitalization mechanism. This concept has originally been proposed by C. Alexander in 1987 in buildings architectural domain. Several years later, this concept has been adapted to other domains. Some of the early pattern work, mostly concerning human interfaces, was done by K Beck and W. In the 1990s, classifications of design experiment has been made to define and create design patterns (Gamma, 1995). Other experts have been closely involved in these efforts such as P. Coad, D. D'Souza, N. Kerth and B. Anderson (Larsen, 1999). In this section, one pattern used in the EMC Repository meta-model, is presented: Composite design pattern.

4.1. COMPOSITE DESIGN PATTERN

The "composite" design pattern (Gamma, 1995) organizes objects into tree structures to represent composite/component hierarchies. It allows to consider individual objects and compositions uniformly.

The composite pattern defines class hierarchies consisting of leaf objects and composite objects. Composite objects are treated the same way as leaf objects. In Our work, a process (see section 5) is made of sub-processes (SP1, SP2, ...) and activities (A1, A2, A3, ...) (figure 3).

Activity functions are used to execute activities. So each activity is made of many functional operations, which are atoms of functionality.

The UML class diagram (figure 4) illustrates aggregations between processes, sub-processes, activities and functional operations. Such a diagram shows the hierarchy defined in figure 3.

A process or a sub-process is a composition of activities (figure 3). In other terms, an activity can be a generalization of a process, which is itself a specialization of some activities (figure 5).

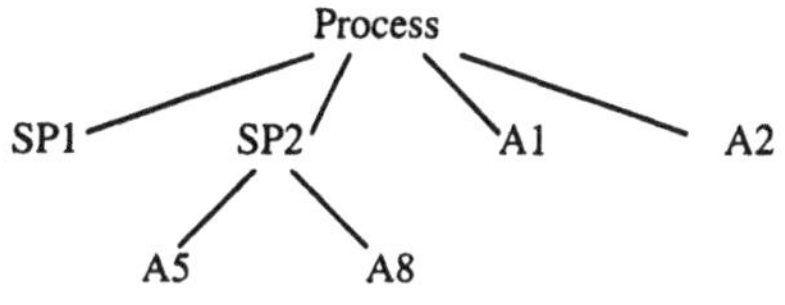

Figure 3 – A process tree structure

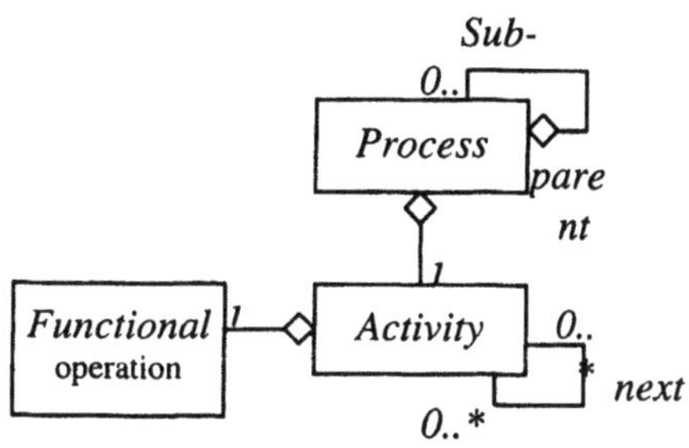

Figure 4– A UML class diagram showing a process tree structure

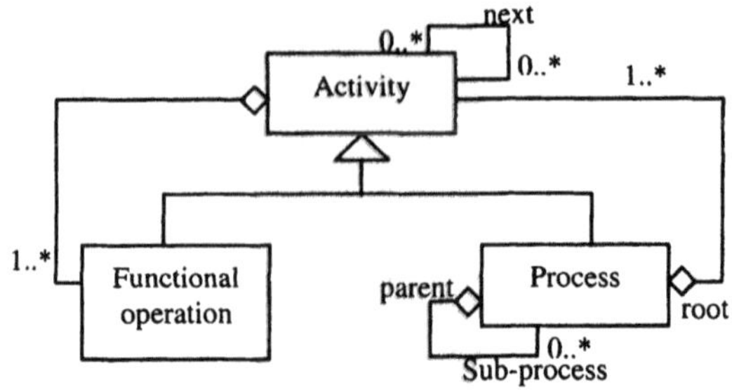

Figure 5 - A process tree structure and heritage between activity, process and functional operation

Other patterns (such as the "Role" pattern (Coad, 1992)) have been used to improve the CIMOSA-based meta-model that is implemented in the EMC Repository.

The next section presents such a meta-model and defines some of its constructs.

5. EMC REPOSITORY META-MODEL

The EMC Repository meta-model (figure 6) is CIMOSA-based. It allows to manage and design business processes in enterprise.

This meta-model covers functional (processes and activities), informational (enterprise objects, object views), resource (functional entities, capability set) and organizational (organization cells, actors, roles) views (Vernadat, 1996).
Its static structure is represented by an UML (Unified Modeling Language) class diagram (figure 6). It allows to enterprise modelers and designers to develop a particular model by using CIMOSA approach.

The figure 6 depicts a DOMAIN meta-class, which is composed of domain processes. Domain processes are triggered by external/internal events. Domain process behavior is described by behavior rules. Execution of these rules permits business processes to be triggered.
Each business process is decomposed into activities, which run according to business process behavior.
Activities use specific resources (functional entities) to activate functional operations, which are elementary functionalities of enterprise.

The principles meta-model classes are more described as follow:

Domain: domains are modules that encapsulate enterprise models, used to break down system complexity. They interact with one another by the exchange of events and object views (e.g. data, documents, results, ...)

Process: A process is a partially ordered set of steps. Process steps can be sub-processes or activities (i.e. elementary steps). Processes are triggered by one or more events. A process P can be formally defined as:

$$P = \langle \text{Pid, alphabet, triggering-condition, behavior}$$

Where *Pid* is the process identification, *alphabet* is the set of process steps, *triggering-condition* is a Boolean expression made of triggering events which needs to be true to start the process and *behavior* defines the control flow (or workflow) of the process. In our work, business processes are modeled. These ones are gathered into domain processes.

Enterprise Activity: an activity is the locus of action. It transforms inputs into outputs over time by means of resources. An activity A can be formally defined as:

$$A = \langle \text{Aid, pre-cond, input, output, req-roles, post-cond} \rangle$$

Where *Aid* is the activity identification, *pre-cond* is a set of pre-conditions to be satisfied to enable activity execution, *input* defines the input object flow, *output* defines the output object flow, *req-roles* indicates the role(s) to be played by the executing resource(s) and *post-cond* defines the set of post-conditions (for instance, ending statuses as found in CIMOSA). To capture more of the semantics of activities, input and output flows can be categorized as control, function and resource flows. The activity behavior is defined by its functional operations.

Event: an event depicts a change in the system state. It represents a solicited or unsolicited fact, which will trigger a process. For instance, the arrivals of a customer order, a machine failure. An event Ev can be formally defined as:

$$Ev = \langle \text{Evid, predicate, time, related object} \rangle$$

Where *Evid* is the event identification, *predicate* is a Boolean expression which evaluates to true when the event becomes active, *time* is clock time when the event occurs and *related object* is an enterprise object associated to the event (for instance, the customer order in the case of a customer order arrival).

Resource: This is a special class of enterprise objects used in support to the execution of activities. As such, it inherits all properties of Enterprise Object, but in addition it gives the list of roles that the resource can hold, the resource availability (usually in the form of a calendar) and the resource capacity. A usage cost per unit of time can also be given. Three generic classes of resources, from which all other types can be derived, can be defined as suggested by CIMOSA: IT applications, humans and machines.

Enterprise Object: An enterprise object is any entity (product, order, resource) which is used, processed, transformed or created by activities in the day-to-day operations of the enterprise. Enterprise objects and their states (object views) are the elements involved in the input/output flows of activities. Enterprise objects are defined by their properties, i.e. attributes for static properties and methods for behavioral properties. An enterprise object EO can be formally defined as:

$$EO = \langle \text{EOid, isa, partof, properties} \rangle$$

Where *Eoid* is the enterprise object identification, *isa* indicates if the object is a sub-type of a more generic object, *partof* indicates if the object is a component of a compound object and properties gives the list of properties describing this enterprise object class.

Organization Unit: An organization unit defines an element of an organization structure provided with authority and responsibility on identified activities and enterprise objects of the enterprise. It defines a decision center at a certain decision level (position, department, division, direction, ...). An organization unit OU can be formally defined as:

OU = ⟨OUid, responsible, responsibilities, authorities, horizon, period⟩

Where *Ouid* is the organization unit identification, *responsible* is the person in charge of this organization unit, *responsibilities* is the set of responsibilities assigned to the organization unit, *authorities* is the set of authorities exercised by the organization unit, *horizon* is the time-frame on which decisions are made and *period* is the time-scale after which decisions can be revised.

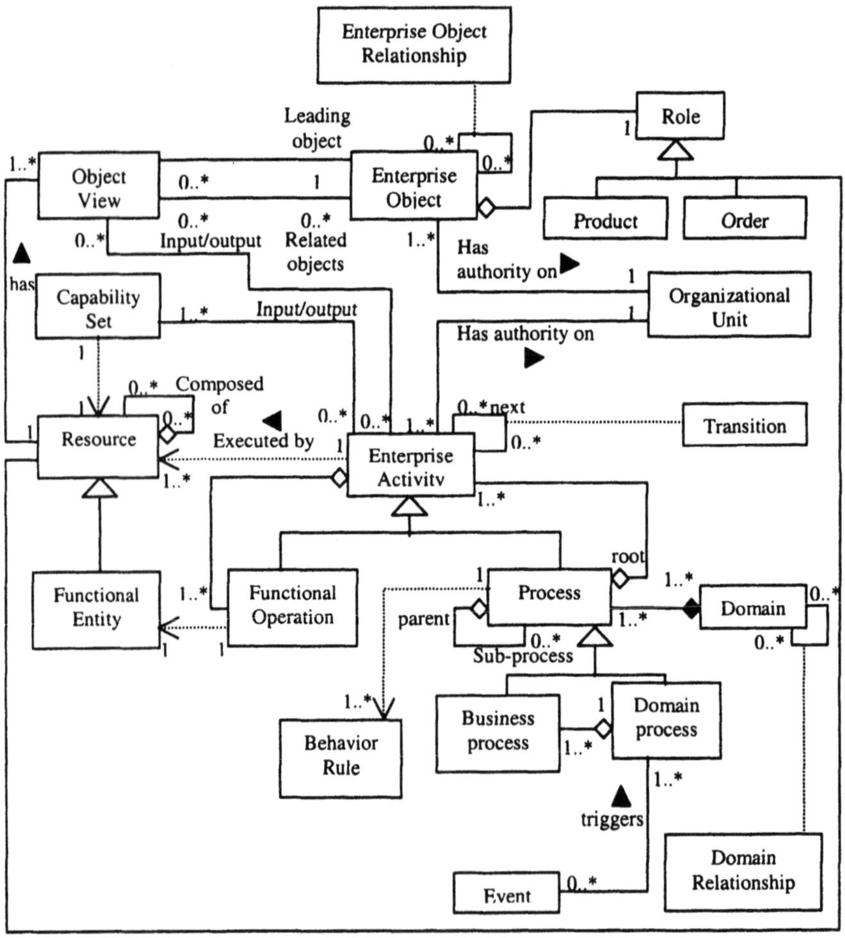

Figure 6- A global (partial) view of the EMC Repository meta-model

6. PERSPECTIVES AND CONCLUSIONS

The proposal of an EMC Repository based on CIMOSA concepts offers better re-usability for enterprise models. However, the problem is that there are several enterprise modeling languages currently in use in enterprises and they should not be ignored (Vernadat, 1999). These existing languages gather enterprise knowledge and know-how. The proposed repository is aimed at playing the role of a components base allowing interoperability with various modeling languages and application systems (Fingar, 2000).
CIMOSA has been selected because it is a reference architecture, which covers modeling and integration. It includes four core views of enterprises (function, information, resource and organization views). It intervenes after the strategic level and covers the CIM (Computer-Integrated Manufacturing) system life cycle (Singh, 1997).

Patterns are used to facilitate re-usability and to capture more of the design knowledge and know-how. UML is a high-performance unified language for modeling, thanks to its static and dynamic diagrams.
Future work will consist in defining constraints for the Repository meta-model using OCL (Object Constraint Language) (Roques, 2000). OCL allows expression of formal constraints, and complements UML diagrams. The next step is to define components from the meta-models, and to implement them with XML (eXtensible Markup Language), which is a standard for interoperability. In fact, developing WEB Services by XML components permits services interoperability (MSDN, 2001).

7. REFERENCES

1. AMICE, CIMOSA: Open System Architecture for CIM, 2nd edition, Springer-Verlag, 1993
2. Booch G., « UML in Action », Communications of the ACM, October 1999, vol42, N°10.
3. Coad P., « Object-Oriented patterns », Communications of the ACM, September 1992, vol35, N°9.
4. CIMOSA Technical baseline, CIMOSA Association, July 1994.
5. Eriksson H.E., Perker M., « Business Modeling with UML-*Business patterns at work* », Ed.Wiley, 2000.
6. Fairchild A., «Interoperability for Enterprise Information Systems», ed.Computer Technology Research Corporation, December 1996.
7. Fingar P., « Components-based frameworks for E-Commerce », Communications of the ACM, October 2000, vol42, N°10.
8. Gamma E., Helm R., Johnson R., Vlissides J., « Design patterns elements of reusable object-oriented », Ed. Addison-Wesley, 1995.
9. Gzara L., « Les patterns pour l'ingénierie des systèmes d'information produit », Ph.D. of *Institut National Polytechnique de Grenoble*, December 2000.
10. Larsen G., « Designing component-based frameworks using patterns in the UML », Communications of the ACM, October 1999, vol42, N°10.
11. OMG, «UML Resources Page», http://www.omg.org/technology/uml, 2001.
12. MSDN, «XML Web Services», http://msdn.microsoft.com/library/, 2001.
13. Roques P., Vallée F., « UML en Action-de l'analyse des besoins à la conception en Java », Ed. VALtech, 2000.
14. Singh V., « The CIM Debacle methodologies to facilitate Software Interoperability », Ed. Springer, 1997.

15. Vernadat F.B., « Enterprise Modeling and Integration-principles and applications », Ed. Chapman & Hall, 1996.
16. Vernadat F.B., « Techniques de modélisation en entreprise : applications aux processus opérationnels », Ed. Economica, 1999.
17. VERNADAT F.B., ABDMOULEH A., BENHALLA S., HEULLEY B.,« Towards a unified language for enterprise modeling : UEML », Engineering & Design Automation (EDA 2001) conference in Las Vegas, august 2001.

GEOGRAPHICAL INFORMATION PLATFORM FOR DISTRIBUTED DATA ACQUISITION SYSTEMS

Nuno Pereira, João Valente, Arminda Guerra
Escola Superior de Tecnologia, Instituto Politécnico de Castelo Branco,
Avenida do Empresário, 6000 Castelo Branco, Portugal.
Tel: +351 272 339300, nunomsp@est.ipcb.pt; valente@est.ipcb.pt; aglopes@est.ipcb.pt

Nowadays are countless the situations where it is necessary the use of a distributed data acquisition system, being sometimes necessary to know the exact location of remote data acquisition units and to integrate more than one type of communication networks.
We have developed a generic geo-referenced data acquisition platform that integrates, in vectorized maps, position and sensor information. A communications module allows using Ethernet, GSM and GPRS networks. It is also opened the possibility to use UMTS.
The first developed application supported by this platform is a robotized system for the measurement of perimeters. Once the remote units can have several sensors, this platform allows registering not only the position of the unit but also all the associated sensorial information.

KEYWORDS: Geographical Information Platform; Robotised System; GPRS

INTRODUCTION

We live in the so-called "Information Era", in which we want to have access to a wide range of information (text, image, sound, video...) at any time and anywhere. Being profit one of the main running motors of technological evolution in the communications area, the available means of communication were designed mainly to serve population masses. The application of these technologies to industry is much slower. The massive use of Internet made it possible to find Ethernet almost anywhere. This availability leads to the development of new protocols for industrial data acquisition networks.

The use of Ethernet in industry is growing. Although for remote data acquisition systems, where we do not have the physical connection to an Ethernet network, there aren't good solutions in the market. Actually the market solutions for wireless systems are or extremely expensive. For small bandwidth, we can use GSM (9600bits/s), or recently, the GPRS that can arrive to (171.2kbits/s) (Simon, 1999). The mainly difference between these two solutions, in terms of service, is in fact, in GSM network the taxation in done according the time connection, while in GPRS the taxation is done by the amount of transferred data. The GPRS solution for less exigent systems in bandwidth terms is a good solution. For higher rates the third generation networks (UMTS) promise to present a permanent connection with a bandwidth of 2Mbits/s.

In this work we present a geo-referenced platform for distributed data acquisition systems. This platform integrates a communication module that besides using TCP/IP, GSM and GPRS, also allows inserting new communication technologies.

APPLICATION DESCRIPTION

This article refers to the development of a **Geographical Information Platform for Distributed Data Acquisition Systems**. This platform receives from data acquisition units the information concerning its location in a certain space, as well as information of any sensor that the unit might have. The information will be stored in a database and visualized in maps or plants.

This application supports any sensor independently of the number of parameters used to send the information, without having to proceed to any alteration in the conception. This application is versatile from the point of view of offering no restrictions related to the type of data or number of parameters necessary to store the information of a certain sensor (for example, an air temperature sensor just needs to send one parameter, the measured temperature in a certain instant, but a position sensor needs to send three parameters: latitude, longitude and altitude).

Summarizing, this application makes it easy for the user to visualise the position of the remote data acquisition units in a map, and allows consulting the received information from each sensor. It also has a module that allows communication through Ethernet, GSM and GPRS, and in the future UMTS.

Architecture

There are three modules (see figure 1) involved in the architecture of this platform:

- The processing module;
- The database;
- The communications module.

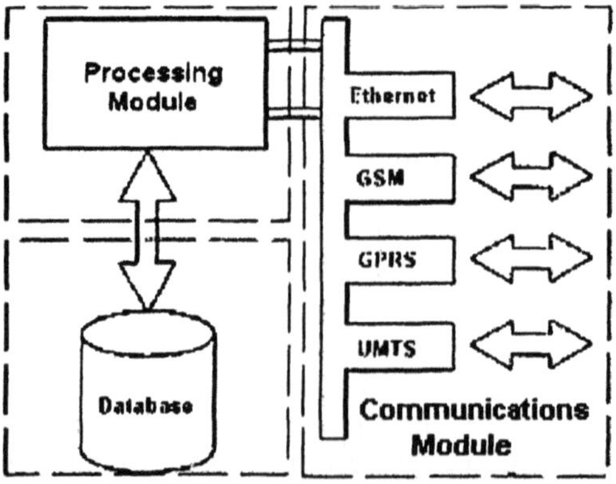

Figure 1 – System Architecture

The processing module is responsible for the treatment of the information that enters in the system. It is here that the information sent by the remote acquisition data units is analysed and registered in the database, as well as maps are updated if necessary.

The database is used to store information sent by the sensors, as well as to maintain information on the several entities registered in the system.

It is the communications module that allows establishing communication between the application and the data acquisition units. Presently, in the implemented application, it is possible to communicate between the application and the data acquisition units through GSM or Ethernet using TCP/IP protocol or GPRS. Communications Module is in itself a modular entity once that it integrates several modules that allow the communication through a specific road, and makes it easy to insert other communication technologies as, for example UMTS.

Practical Application

As an application example of this platform we have implemented a Robotized System for Measuring Perimeters (RSMP). The RSMP receives the robot's position information and draw's their trajectory in a map or plant, as shown in figure 2. This application will then calculate the areas and perimeters of the trajectories made by the robots. The information obtained by the sensor(s) placed in robots is stored in the database as well as the information regarding their position.

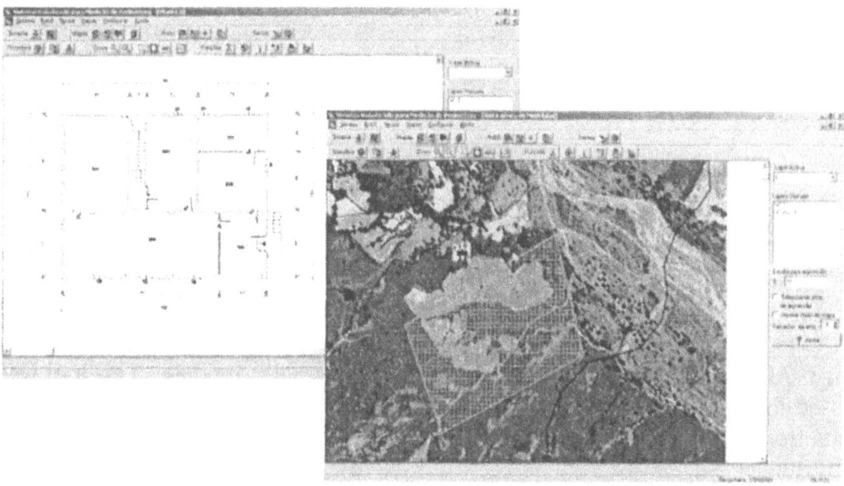

Figure 2 – Use of the Robotized System for Measuring Perimeters in two different situations: in the plant of a building *(left)* and in a land parcel *(right)*.

From the functional point of view, three different modules characterize the SRMP application:

- The information control module, that processes and treats the data that enter into the system;
- The communications module, that implements the necessary protocols to communicate with the exterior;
- The graphic module, which makes the user interface and allows the visualization of the system through maps.

The interactions between the modules are illustrated in figure 3. When a message is sent by a remote unit (robot) the communications module receives it, treats it and delivers the data to the information control module. This module analyses the information; if the data is coming from a sensor it makes its registry in the database; if it is position information it is also given to the graphic module. This module checks if the corresponding map is opened and refreshes it.

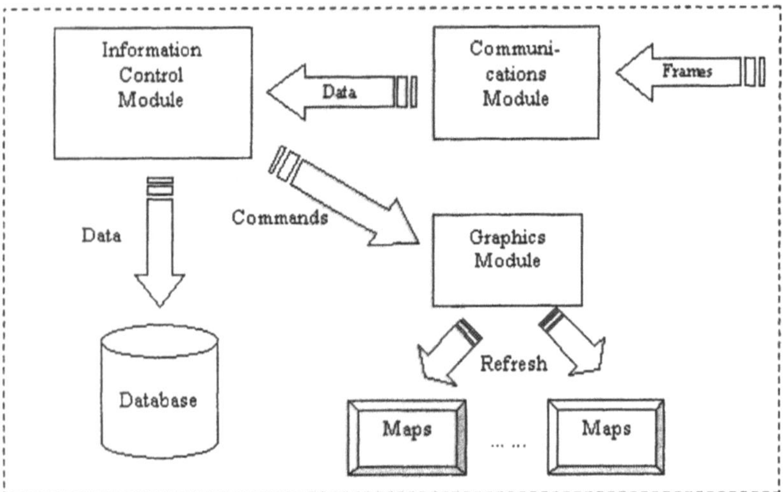

Figure 3 – Interactions between different modules

Used Technologies

Some computer tools were used to develop this application. The DbCAD dev package (ABACO, 1999) was used to implement the graphic module. This tool was also used to develop the geographic database. Microsoft Access was used to store the data from sensors and other parameters. Object guided programming in C++ was used in Builder 5.0.

CONCLUSION

The developed application facilitates the reception of information acquired by remote acquisition data units, which sends the information according to a specifically designed protocol. The object orientated programming model, used in this application, makes it flexible because it facilitates the insertion of new modules that will add new functionalities to the program, as for example the support for other communication technologies.

The communication modules already implemented are GSM, GPRS and Ethernet. The development of technology, namely with the insertion of a stack TCP/IP in a great number of microprocessors, allows to receive information from sensors without using a PC. Thus, it becomes possible to connect to Ethernet a great number of devices, including wireless devices, which increase the connection possibilities between the supervision unit and the remote units, independently of their location.

The use of vectorized maps allows to overview the location of the remote units, either mobile or fix.

The protocol designed for this application doesn't offer restrictions to the type of data, making this application quite flexible in terms of using new sensors in the remote units. The possibility to insert and configure new maps, units and sensors turns this platform into a tool that is easily adapted to a great number of configurations of distributed acquisition data systems, not only at the supervision but also at the control level.

FUTURE WORK

In the future, the support for communication with USB devices and UMTS technologies will be implemented.

Once the developed protocol is not compatible with the norm NMEA 0183 (norm of GPS) defined by National Maritime Electronics Association, the entity that normalises the communication among GPS devices, it will be developed a module that will make the conversion between the protocol defined by the referred norm and the protocol used by this application.

BIBLIOGRAPHY

1. ABACO. DbCADdev – User and Reference Guide. ABACO srl, 1999.
2. Booch G, Rumbaugh J, Jacobson I. The Unified Modeling Language - User Guide. Addison-Wesley, 1998.
3. Ezzel, Ben. Mastering in Windows 2000 - Programming with Visual C++. SYBEX 2000.
4. Simon Buckingham. Data on GPRS. Mobil lifestreams 1999.

DEVELOPING GRAIXPERT: A SOFTWARE TOOL TO ANALYSE AND DESIGN PRODUCTION SYSTEMS

Paul Eric Dossou[1], David Chen[1], Roger Bertin[1] and Guy Doumeingts[1,2]
[1]*LAP/GRAI, UMR CNRS 5131, University Bordeaux 1*
351 Cours de la Libération, 33405 Talence FRANCE
{Dossou, Chen, Bertin}@lap.u-bordeaux.fr
[2]*GARISOFT*
Parc d'Activité FAVARD, 33170 Gradignan, France
gdoumeingts@graisoft.com

This paper presents the latest development of a computer aided design tool 'GRAIXpert'. It aims at supporting the use of the GRAI methodology for the analysis and design of production systems. At first the global architecture of GRAIXpert is introduced. Then the design process and its associated problem solving method are discussed in detail. A simplified example is given to illustrate the proposed approach. Perspective and future work are considered as part of the conclusion.

Keywords: Design, Production system, Enterprise modelling, CAD

1. INTRODUCTION

The GRAI Method (Doumeingts, 1984) and the GRAI Integrated Method (GIM) extended to the GRAI Methodology (Doumeingts, 1999), aim at analysing and designing various sub-systems of the production system. The GRAI approach consists of four phases: (1) An initialisation phase to start the study, (2) A modelling phase to model the existing system, (3) An analysis phase to detect the inconsistencies of the studied system, and (4) A design phase during which the inconsistencies detected are corrected, and a new system specified. The approach presented in the paper is limited to the decision sub-system and to the design phase.

Within the frame of GRAI methodology, a computer aided modelling/analysis and design tool is being developed: e-MAGIM. In this domain, several research works have already been carried out (Zanettin, 1994), (Akif, 1995), (Durcos, 1995), (Moudden, 1997). This paper presents the latest improvement of the e-MAGIM tool through the development of a new module: GRAIXpert (Dossou, 2000). It is a hybrid expert system which interacts with the human expertise (knowledge and experiences of experts of GRAI methodology) and supports the both analysis and design phases. The tool aims to:

- capitalise knowledge in production system engineering, which is usually dispersed;
- shorten the delay needed to design (or redesign) production systems;

- support the systematic use of enterprise engineering methodologies;
- promote computer-based enterprise modelling and design; and finally
- improve the performance of production system.

This paper will firstly present the architecture of GRAIXpert with a brief description of its sub-modules. The classification of the rules used during the analysis and design phases is refined. Then a design process is proposed, which integrates the various modules already used in e-MAGIM. Finally the Problem Solving Method associated to the hybrid expert system is presented. It allows combining different reasoning techniques such as CBR (Case-Based Reasoning) Decomposition reasoning, etc. A simplified illustration example to design decisional sub-system will be shown at the end.

2. GRAIXPERT: THE ARCHITECTURE

The design of hybrid architectures for the intelligent systems is a new domain of AI research for the development of the future intelligent systems. The intelligent systems represent a combination of different techniques like fuzzy logic, neurone networks, genetic algorithms, etc. (Yahia *et al.*, 2000). They are also concerned with the integration of these techniques to the conventional systems and the statistic methods, the CBR for example. The architecture proposed takes into account the expert's knowledge in terms of reference models and rules resulting from the expert knowledge acquisition. The e-MAGIM Kernel is used for modelling the existing system. It is a graphic editor to elaborate the representation of various models of the GRAI methodology. The architecture is composed of three sub-modules: the Manager, the Knowledge Capitalisation and the Knowledge Based System (see figure 1).

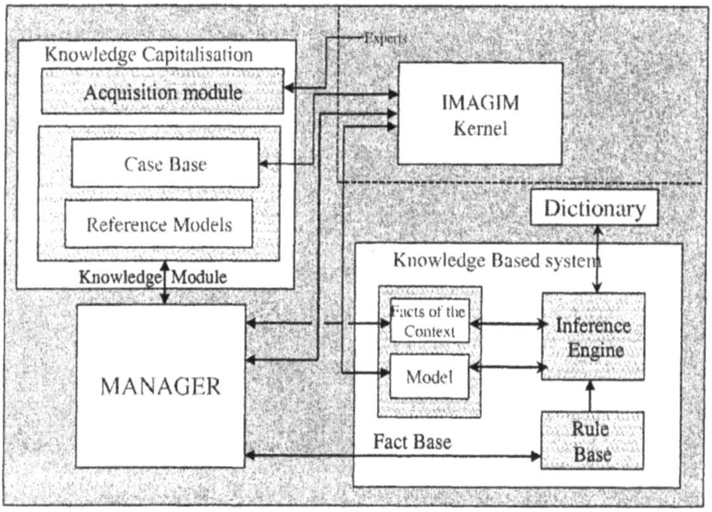

Figure 1 - Architecture of GRAIXpert

2.1 The Manager

The Manager controls and manages the system's interactions with the users. It presents to the users with some appropriated questions and choices, the necessary information about the characteristic of the studied enterprise. It also manages the rules classified according to a typology of production systems. Its main tasks are modifying, suppression, or selection of the applicable rules in a given context. It is also used for the loading and the saving of rules files. Finally, it controls the design process, different actions of the sub-modules and their interactions.

2.2 The Knowledge Capitalisation

The Knowledge Capitalisation is composed of reference models associated to a domain and of previous studies. The knowledge capitalisation process needs some aptitudes to manage different know-how and points of view. It must integrate this knowledge in an accessible, usable and maintainable form. It is proposed an expertise model based on the knowledge of the experts but also on the previously realised studies.

A reference model is a generic model for a given domain so that a design solution is a subset or a particularisation of this model. It must be put in a form that is able to be manipulated by the design process. To elaborate a reference model, a typology of production systems must firstly be defined. There exist a certain number of classifications in the literature but none of them is suitable for our study. The classification used in the paper is based on two criteria: (1) type of production such as *Continuous production*, *Discrete Production*, *Service*, etc. (2) *repetitiveness of production*. For example, the discrete production can be sub-divided into: *Mass production*, *Small and medium production*, and *One-of-a-kind production*. For each class of production systems identified above a decisional reference model, a functional reference model and a process reference model must be elaborated. The present paper focuses on the Small and Medium size production systems and decisional model.

The decisional reference model is elaborated using GRAI grid formalism (Doumeingts, 1984, 1999). Decision levels, decision functions, decision centres and links (decisional and informational) between decision centres are defined. For each decision centre, a decision activity reference model is also defined using GRAI nets formalism (Doumeingts, 1984, 1999). This paper only deals with the global decisional reference model (GRAI grid reference model).

The knowledge capitalisation module contains two sub modules:
- The acquisition one is used to improve the acquired knowledge. It serves to modify, add, and suppress the knowledge in the knowledge sub-module.
- The knowledge sub-module with two databases. The first is the case base where previous studies are stored. The second is composed of reference models.

2.3 The Knowledge Based System

The Knowledge Based System contains a rule base, a fact base to store the existing system's models obtained with e-MAGIM Kernel, but also the fact base of the context containing the data of the context and corresponding to used facts in the rule base, and an inference engine. A dictionary is used to translate user's expressions into standard expressions provided by the GRAI methodology.

The GRAI methodology elaborates various models with the help of the formalisms. So the rules established have to be classified according to the formalism, the model, the integration between them, and the intra-model or inter-model consistency. The main criteria of classifying rules are related to: (1) the role of the rules and when they are used; (2) the model and its environment; and (3) the formalism used. According to the first main criterion, four types of rules are defined: (1) The model construction rules, describing the consistent use of the GRAI formalisms during the modelling phase. (2) The model verifying rules for the first level of diagnosis. They are used at the end of the modelling phase and the beginning of the analysis one. (3) The system verifying rules representing the real diagnosis rules. They are used during the analysis phase. (4) And the design rules to help the user during the design phase of the GRAI methodology. The second main criterion of this classification divides each previous type of rules into groups. They are related to: (1) a diagram; (2) a view or a sub-system; and (3) a model. The third main criterion is related to the formalism and divides each group into categories. For each group, we obtain some rules related to the formalism: (1) Actigramme (functional and physical models); (2) GRAI grid (global decisional model); (3) GRAI nets (local decisional model); (4) Extended Actigramme (process model); and (5) Entity/Relation (informational model). There is a hierarchy between the three main criteria. Each class obtained by the decomposition of the first main criterion is decomposed in following the second main criterion. Then each result is also divided by following the third main criterion.

3. THE DESIGN PROCESS

A design process is elaborated using the knowledge of experts of GRAI methodology, and the approach they follow during the design phase. This process has five sub-phases corresponding respectively to the design of each new model of GRAI methodology. The first sub-phase is functional view design; the second is physical system design. The third, "decisional system design" allows to specify new decision system models (global and local). The other two sub-phases are process view and information system designs. Only the decision system design will be presented in the paper.

3.1 New approach of preliminary design

The basic concepts of this approach are derived from some design theories (Yoshikawa, 1981; Tomiyama, 1995) and (Chen, 2001). An artefact is specified by a set of property descriptions $P=\{p_1,p_2,\ldots..p_n\}$ and a set of structure descriptions $S=\{s_1,s_2,\ldots s_n\}$ which determines the properties. The design knowledge is represented by the classifying concepts on a set of known artefacts $A=\{a_1, a_2,\ldots a_n\}$. The design is an iterative process that transforms an initial state of an artefact specifications $\{Q_0\}$ into a final state $\{Q_n\}$ representing the complete specifications of the designed artefact. The state Q_i obtained by the transformation stage *i* is represented as:

$$Q_i = \{P_i^+, P_i^-, S_i^+, S_i^-\}$$

with P_i^+ the set of the desired properties satisfied by previous transformations and P_i^- those to be satisfied, S_i^+ the set of validated structure descriptions providing the set of property descriptions P_i^+, S_i^- those which are not yet validated.

Among possible design transformations, three of them are considered as fundamental: Synthesis (T_S), Analysis (T_A) and Refinement (T_R). A design transformation at each iteration *i* is performed by one of the three transformations: $f = < T_S, T_A, T_R >$ as illustrated in figure 2.

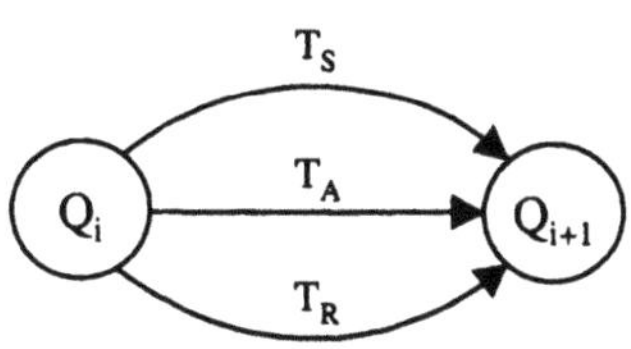

Figure 2 - The three basic transformation processes of designing

The *synthesis* process T_S maps desired property descriptions (P) onto structure descriptions (S). The *analysis* process T_A is the inverse process of synthesis. It maps structure descriptions (S) onto property descriptions (P). The *Refinement* T_R is a process of detailing and decomposition of either property (P) or structure (S) descriptions (see figure 3).

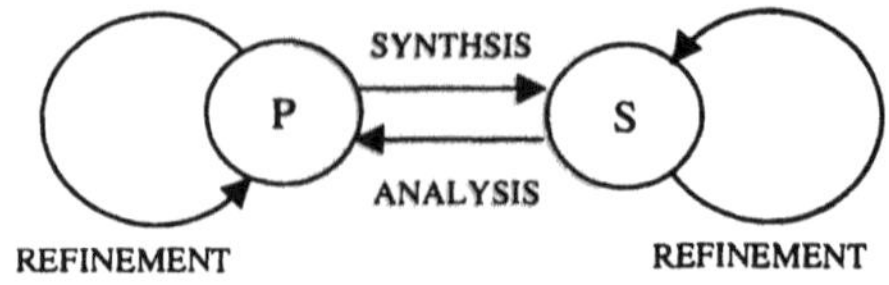

Figure 3 - Relations between the three transformation processes

3.2 Design process of the decision system

The design of the decision system consists of five macro-steps; the first four steps allow elaborating the global decisional model and the last one the local decisional models. For this sub-phase we have also established some reference GRAI grid models taking into account some criteria like activity domain and production repetitiveness in order to classify these reference models. Some reference GRAI nets models for each decision centre of each reference grid are also defined. The decision system design starts with the results of analysis phase and the design objectives. These macro-steps are: (1) choice of the reference grid model, (2) evaluation, (3) preliminary design, (4) detail design for the global decision model, and (5) GRAI nets design for the local decision models.
This design process combines different types of reasoning and integrates RBR (Rule Based Reasoning), MBR (Model Based Reasoning) and CBR (Case Based Reasoning). The main reasoning is the generalisation and classification where other reasoning like CBR, the Decomposition Reasoning, the reasoning of direct corresponding, and the transformation reasoning are also envisaged.

4. PROBLEM SOLVING METHOD

For designing a new production system, one starts with the functional model, followed by the physical model. Then global and local decision models are elaborated and information model specified. The problem solving process used for the functional, process, or decisional models is different from physical or informational one. For example, when designing the functional model, one firstly uses a generalisation reasoning to get a functional reference model, and then the transformation reasoning to adapt this reference model in order to obtain the new functional model.

4.1 The CBR Mechanism

Each reference model is considered as an object with a structure. The CBR mechanism (see figure 4) uses an algorithm «extract - select - transform » to choose the reference model that is suitable for the system studied. In effect there are in the case base different types of objects. One must firstly choose the object class C_i on which the mechanism will be applied. Each class has different objects:

$$C_i = \{O_{i1}, O_{i2}, \ldots O_{in}\}$$

The extractor allows the choice of the reference class. The selector chooses the reference model suitable to the study. An object O_{ij} of the class C_i is chosen by comparing the parameters of the object to design O_m, result of the modelling/analysis phase and the design objectives, with the objects O_{ij} of the class C_i. Then the object O_{ij} is used for realising the design of the object O_m. The transformer (according to the manager) manages the process with which one obtains the reference model and the context which is associated to it. It uses firstly the decomposition reasoning and the direct corresponding reasoning or secondly using transformation reasoning. The result of the CBR mechanism is the new design object O_c (new functional model). The transformer also manages the possible modifications of the reference model which permits to get the preliminary design model (preliminary functional model or preliminary GRAI grid model).

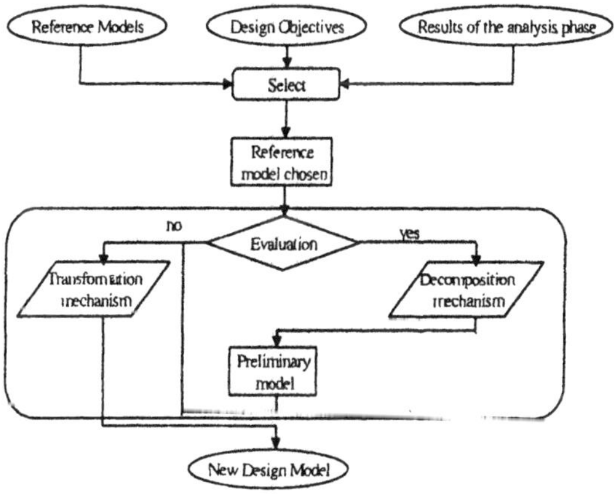

Figure 4 - CBR mechanism

4.2 The Decomposition Mechanism

In effect, the decomposition mechanism is used on an object O_{ij}. Each object O_{ij} is composed of a set of sub-objects $O_{ij}^{\ k}$. We obtain:

$$O_{ij} = \bigcup_{k=1}^{n} O_{ij}^{k} \quad \text{with n equal to the number of sub-objects.}$$

For example, one can decompose a reference grid into decision centres according to the GRAI grid formalism. Then, each object $O_{ij}^{\ k}$ is studied separately. The design of this object is realised according to the design objectives. One transforms by iterations the object $O_{ij}^{\ k}$ into the design object O_{ck}. The design solution of the object O_{ij} is obtained by unifying the different partial solutions of the objects $O_{ij}^{\ k}$.

$$O_c = \bigcup_{k=1}^{n} O_{ck}$$

We define a linear application L from O_{ij} to O_c. Each element $O_{ij}{}^{k}$ of O_{ij} is associated to the solution element O_{ck}.

$$L: \quad O_{ij} \longrightarrow O_c$$
$$O_{ij}{}^{k} \longmapsto O_{ck}$$

This application L expresses the design of the new object. Then, the re-composition allows the association of all the designed objects. Finally, a new global object (structure) can be obtained. For example, one can design each decision centre of the preliminary grid (solution) by direct corresponding to obtain some partial solutions. Then one recomposes the structure of the grid. The algorithm « decompose - solve - recompose » allows to get a preliminary design model. This preliminary model is the solution of re-composition of the decomposition reasoning (the adapted solution for the CBR).

4.3 The Transformation Mechanism

The transformation reasoning (figure 5) is used to adapt a preliminary model with the synthesis group/designer in order to get a final model. The design rules of the expert system are the transformation rules. They are combined with the preliminary grid and the results of the analysis/modelling phase to generate some design alternatives. The synthesis group and the designer choose the solution they judge the best, which becomes the solution of the transformation reasoning, the final model.

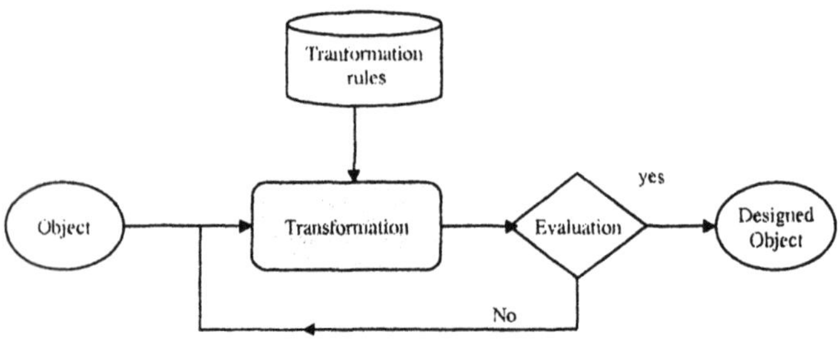

Figure 5 - Transformation mechanism

4.4 The new problem solving method

The problem solving method proposed combines different types of reasoning approaches as shown in figure 6. The main reasoning is the generalisation and classification during which other reasoning techniques such as Case-Based Reasoning (CBR), the Decomposition Reasoning, the reasoning of direct corresponding, and the transformation reasoning are also used.

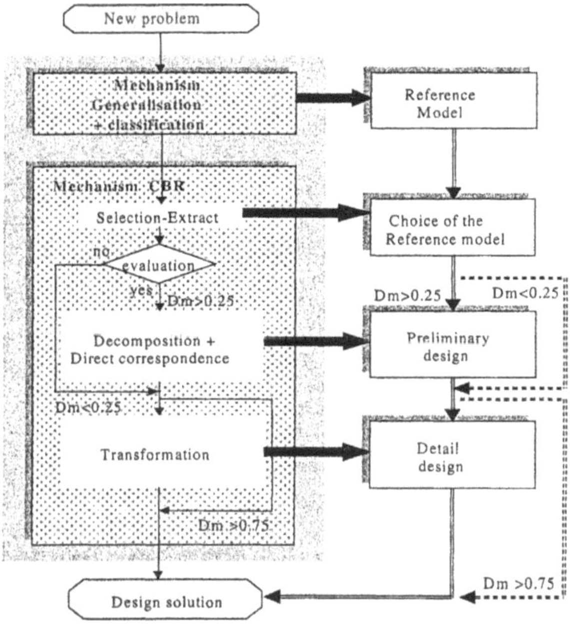

Figure 6 – Relations between reasoning mechanisms and design process

For each macro step one mechanism is used. The CBR mechanism uses the decomposition or the transformation mechanism for its adaptation phase. It is the combination of the three mechanisms used for the design of the functional, process or decisional models. One uses transformation reasoning to design the new information or physical models. The CBR is used to design the new GRAI nets with the help of design cases (the reference GRAI nets previously established). The first two parts of the CBR algorithm (extract - select) allow choosing a reference GRAI net and the third part of the algorithm (transform) to generate design alternatives.

5. ILLUSTRATION EXAMPLE

The example is largely simplified to illustrate the proposed approach to design a decision system. Prior to the study, it is necessary to formalise the reference model by describing property and structure attributes of all its decision centres. In other words, each decision centre of the reference GRAI grid model is described by a set of property attributes (functions, roles,...) and structure attributes (specifying its location in the GRAI grid). Design in this case is to transform an initial state containing required property descriptions (user requirements, objectives,...) and possibly some structural descriptions (existing decision centres that users wish to keep in the future system) to a final state containing all information to compose a GRAI grid model. This iterative transformation is realised on the reference model description spaces using the three basic transformation processes i.e. Synthesis, Analysis and Refinement.

Following the approach of GRAI methodology, at first e-MAGIM graphical editor was used to elaborate various models of the existing system of studied company. For example, on a GRAI grid model of the existing system, the analysis phase has detected some inconsistencies using the Knowledge-Based system (diagnosis rules). To design a new grid G_d, taking into account the context of the studied enterprise, the results of the analysis and the design objectives, a reference grid model G_r stored in the Capitalisation Knowledge module is chosen. To choose the reference grid, the Manager has used the CBR mechanism. The extracted grid is compared to the analysis grid to determine the modifying degree D_m. Each attribute (decision centre, function, level...) of the reference grid is compared with the analysis grid. If D_m <0.25 then the design goes to the detail design macro-step using the transformation mechanism included in the CBR. Otherwise the design must go through the preliminary design macro-step before the detail one. For the studied case D_m is equal to 0,82, so that the preliminary design macro-step is needed. Based on requirements and results of the analysis phase, a set of designed decision centres are obtained using direct correspondence approach (Dossou, 2000). Then, these partial solutions (decision centres) are re-composed to obtain a preliminary grid. Figure 7 shows the designed GRAI grid model obtained after the detail design (not discussed in the example).

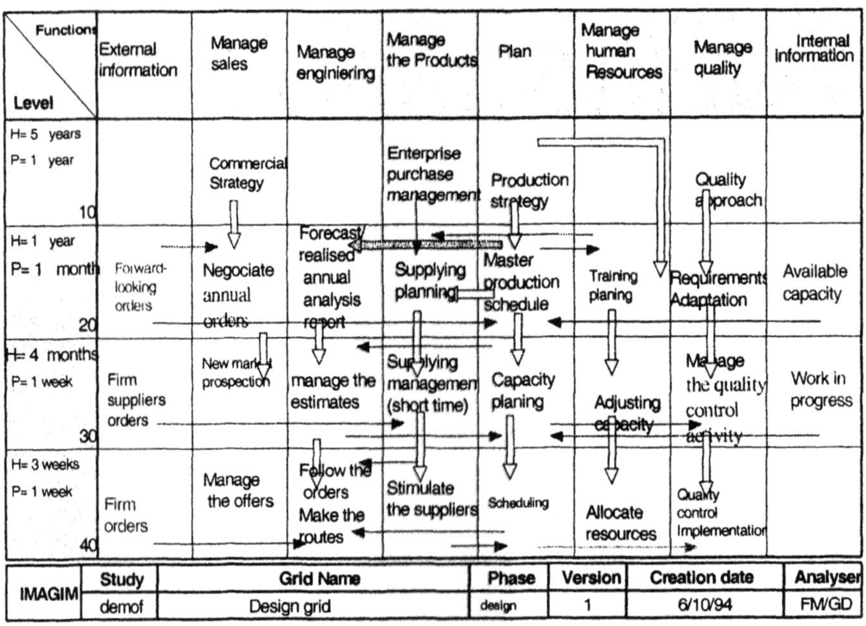

Figure 7 - Designed GRAI grid model of the decision system

6. CONCLUSION

Designing enterprise systems is a very complex task. To deal with such complexity with the help of computer tools, a hybrid problem solving method has been proposed which combines various reasoning techniques. This method is similar to the observed reasoning process of experts of GRAI methodology i.e. it better reflects their experiences when they design production systems using the GRAI methodology. Furthermore the recent development in the area of design theories has been adapted to formalise the design process so that it can be easily implemented in computer. An example dealing with the preliminary design of a decisional system was presented to illustrate the approach.

This tool is being developed by GRAISOFT (a company created to transfer research result to industry). The approach presented in the paper is only part of progressive and continuous improvements to develop e-MAGIM. This computer tool aims at supporting the various phases of the GRAI approach and covers various sub-systems and views. However due to the restriction on the number of pages, only the design phase and the decision system model have been discussed in the paper. It is to note that the modular architecture of the tool allows adding other modules when necessary so that the tool can be enhanced according to new requirements coming from the enterprise engineering and integration practices.

7. REFERENCES

1. Akif. Outil Informatique support d'aide à l'Analyse de Cohérence des Systèmes de Gestion de Production. Thèse de Doctorat, Université Bordeaux I, Octobre 1995.
2. Chen D, Doumeingts G "Towards a Formal understanding of design processes: Contribution to the development of a Theory of Design, In Proceedings of 8th IEEE International Conference on Emerging Technologies and Factory Automation, Juan-Les-Pins, 15-18 October, 2001.
3. Doumeingts G. Méthode GRAI : Méthode de conception des systèmes en productique. Thèse d'état, Automatique, Université Bordeaux I - Novembre 1984.
4. Doumeingts G, Ducq Y. "Enterprise Modelling techniques to improve Enterprise Performances" - The International Conference of Concurrent Enterprising: ICE'99 - March 15/16/17th 1999 - The Hague - The Netherlands.
5. Dossou P, Chen D, Bertin R, and Doumeingts G. "Reasoning modelling for production systems design in a computer aided GRAI environment", In ASI2000 annual conference, September 18-20, 2000, Bordeaux, France.
6. Durcos S. Outils Informatiques supports aux méthodes de Modélisation et de Conception de Systèmes de Production. Thèse de Doctorat, Productique - Université Bordeaux I, 1995.
7. Moudden A. GIMXpert et GIMCase. Une méthode et des outils pour concevoir les systèmes de production. Thèse de Doctorat, Productique - Université Bordeaux I, Octobre 1997.
8. Tomiyama T. A design process model that unifies General Design Theory and Empirical Findings. *In Design Engineering Technical Conferences*, ASME, Vol. 2. 1995.
9. Yahia ME, Mahmod R, Sulaiman N and Ahmad F. 'Rough neural expert system', 'Journal of Expert System with Application' vol 18, issue 2, Februry 2000.
10. Yoshikawa H. General Design Theory and a CAD system. *In Man-Machine communication in CAD/CAM*, T.Sata, E. Warman (editors), North-Holland Publishing Company, 1981.
11. Zanettin M. Contribution à une démarche de conception des Systèmes de production. Thèse de Doctorant, Productique, Université Bordeaux I- Mars 1994.

MANUFACTURING IN THE 21st CENTURY

Vince Thomson
Department of Mechanical Engineering, McGill University
817 Sherbrooke Street West, Montreal, QC, Canada, H3A 2K6
vince.thomson@mcgill.ca

This paper tries to predict the main influences on manufacturing in the 21st century. Given the caveat that no one can tell the future, the main drivers of productivity in the 20th century are considered and extrapolated into the 21st century. Information technology, as the prime driver for future productivity, is discussed, and predictions are made about its influence. The nature of other future technologies is also explored. Finally, the impact of the new trends on society is discussed.

Keywords: 21st century, future, information technology manufacturing, productivity

1. INTRODUCTION

There were many technological improvements that allowed the realization of productivity gains in the 20th century: the electrification of machines, the creation of new materials for better material forming, the adoption of Taylor's "scientific management" for improved processes, the integration of mechanical devices, and the emphasis on quality control. These improvements helped to exploit the division of labour paradigm developed during the Industrial Revolution (Ayers, 1991).

In 1946, Eric Leaver and John Brown wrote the article "Machines Without Men" and promoted the vision of a factory producing goods with no production workers, just automation operated by a "collation" machine, namely, what we call a computer today. In 1973, Joe Harrington wrote *Computer Integrated Manufacturing*, and the integration of information paradigm was born. Driven by computers and miniaturization, information technology is starting to bring about the "Machines Without Men" vision by enabling the integration of information paradigm. As proof of this, great gains in productivity due to information technology did start to occur during the last decade of the 20th century (Conference Board of Canada, 2000).

What about the 21st century? Several groups and authors have written about the influence of and the demand for new technologies at the beginning of this century (Moody and Morley, 1999; Ellyard, 1998; IMS International, 2000; National Research Council, 1998). While the development of new materials and fabrication processes hold the possibility of providing many new capabilities, it is information technology that will have the greatest influence on productivity improvement. Will information technology completely effect the vision put forth by Leaver and Brown? This paper investigates the reasons for past and present levels of productivity; then, the factors which will have the greatest impact in the future, especially information technology, are presented along with predictions of the areas/technologies which will provide the best opportunities for productivity improvement in manufacturing. These include: continued miniaturization of electronic and photonic systems, embedded systems, mass customized software, and the presently elusive integration of information. These technologies will lead to new manufacturing capabilities and greatly improved processes for both fabrication and business. A new era will begin with the systemization of knowledge, the development of 'knowledge' products, as well as the integration of material processing and information systems. Finally, how to manage these technologies for the maximum benefit of corporations and society will be discussed.

2. PRODUCTIVITY

In order to be able to look at the main influences of the 21st century, it is necessary to understand the main drivers of productivity in the past and in the future. In, *An Inquiry into the Nature and Causes of the Wealth of Nations* (1776), Adam Smith wrote: "The greatest improvement in the productive powers of labour ... seem[s] to have been the effects of the division of labour". The Industrial Revolution was born out of the need to provide inexpensive goods for the growing middle class in the 18th century. The division of labour paradigm provided the mechanism to organize labour to be efficient, and distributed power from steam, and later, electricity provided the means to make the paradigm work. Machine tools were later developed to manufacture standard goods in large quantities. The division of labour paradigm persists today and influences every aspect of how we organize activities.

Despite the positive benefits of the division of labour paradigm, starting in 1973, there was a great reduction in productivity growth (Figure 1). Before 1973, productivity growth in most G-7 countries was over 2.5% per year and greater in developing countries. After 1973, productivity growth was about 1% per year in the G-7 countries. What happened? The nature of producing goods had changed. Customers wanted more product variety. Product complexity greatly increased, and so did manufacturing complexity (Ayers, 1991). With the advent of more product variety and complexity, the nature of productivity growth changed, and economy of scale and division of labour paradigms could not cope with the new challenges.

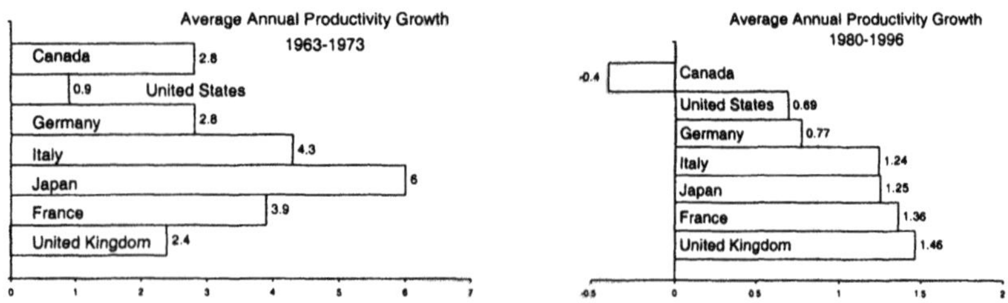

Figure 1 - Comparison of productivity growth for the G-7 countries for 1963-1973 and 1980-1996

Figure 1 - Comparison of productivity growth for the G-7 countries for 1963-1973 and 1980-1996

Table 1 gives a comparison between the main attributes of the Industrial and Information Revolutions. Will the integration of information paradigm be as effective in generating productivity as the division of labour paradigm has been, and more importantly, will integration of information have as pervasive an influence on society as the division of labour? The conclusion of this paper is yes.

Table 1 - Comparison of the industrial and information revolutions

	Industrial Revolution	**Information Revolution**
Period	1700 -1970	1970 -
Currency	material	information
Paradigm	division of labour	integration of information
Enabler	distributed power	information technology
Focus	material flow	information flow

Table 2 - Productivity growth in Canada and the USA (GDP/worker or GDP/hour worked)

Years	**Canada**	**USA**
1989-1994	1.3%	1.2%
1995-1999	1.0%	2.7%

A comparison of productivity growth between the USA and Canada (Table 2) shows that labour productivity growth in Canada and the USA for 1989-1994 was approximately the same, while during 1995-99 it increased to 2.7% per year in the USA and remained about 1% in Canada. The report, *IT and the New Economy* (Conference Board of Canada, 2000), indicates that the reason for the increased productivity in the USA was due to the greater use of information technology. In fact, the report points out that information technology (IT) was responsible for 50% and possibly up to 80% of the total productivity growth in the late 1990's. Why then did the United States lead in productivity growth at the end of the 20th century? It was because of greater technology adoption, faster technology diffusion, and more structural change due to the adoption of IT. In addition, the reports, *Higher Wages Accompany Advanced Technology* (US Dept. of Commerce, 1993) and *Technology-induced wage premiums in Canadian manufacturing plants during the 1980s* (Statistics Canada, 1993), have stated that manufacturers which were better adopters of technology had more market share, achieved higher productivity, and paid better salaries. Employees also had higher education, possessed more skills, and were better trained in these companies.

3. TIME OR PROCESS BASED COMPETITION

Knowledge work has become the most important component in the economy of the industrialized world. Peter Drucker (1993) has stated "that we now see knowledge as the essential resource", and that knowledge worker "productivity, rather than the productivity of the people who make and move things, is the productivity of a developed economy". One way that the emergence of knowledge work will change economies is the shift of focus from product to process. Until the 1980's there was significant differentiation in the world based on companies' ability to make or not make certain products, and this was reflected in the economies of nations, which were (are) based on the competitiveness of these companies. Presently, product knowledge is ubiquitously available, and therefore, airplanes, cars, microelectronics, and telecommunication equipment can be manufactured in any industrialized country. What differentiates companies today, and what will continue to differentiate them, is process knowledge (Haeckel, 1993). The problem of knowledge worker productivity is present in all economies of the world. Shintaro Hori (1993) in an article on Japan's white-collar economy states that a disproportionate growth of overhead is symptomatic of knowledge worker productivity problems. Hori's solution is to focus on white-collar productivity and process improvement. The need to focus on process is reinforced by the economist, Lester Thurow (1992), "in the future, sustainable competitive advantage will depend more on new process technologies and less on new product technologies".

Competing on process not product means focusing on information flow. Improved information flow is achieved with better connectedness and formalized relationships. Computer technology allows control of factory- and business-wide processes for improved responsiveness through integrated information.

One of the key research areas proposed for improving manufacturing has been the organization and systemization of knowledge (Hayashi, 1993). This systemization will allow companies to record and use manufacturing knowledge better for greater flexibility, more rapid response, better control of all processes, and the development of new technology. This systemization of manufacturing knowledge can best be achieved by creating better methods for defining, planning and controlling processes, and not just production processes, but especially manufacturing support processes where most knowledge workers are employed.

If everyone has equal access to knowledge, then, competing on knowledge means competing on time or timeliness, i.e., the speed of turning new knowledge into competitive advantage. Responsiveness to customers is rapid cycle time for new product introduction, order fulfillment and customer service, which are achieved by process flexibility and good management of information. In fact, business performance in the future will be measured by per unit time instead of per unit production (Drucker, 1990). An example of this emphasis on time is the very successful, 6-sigma quality program started at Motorola in 1987. It is based on process renewal where the goal is a 50% reduction in work process time every 12-18 months, where the industry average has been 3-5 years (Blackburn, 1991). In addition, Motorola and others have found that as process time is decreased, quality increases. The latter occurs because, in order to minimize cycle time for product development and production, work must be done correctly the first time.

4. THE FUTURE

4.1 Improvement in Productivity

Change in manufacturing has been highly accelerated in the last half of the 20th century; the Information Revolution will bring about change at an even higher rate in the 21st century. This change will be focused on time based competition, enabled by information technology and managed through concentration on process.

Where will the productivity improvement come from? First, product complexity will increase 15-20 times. This has happened in the 19th and 20th centuries (Ayers, 1991), and thus, should happen in the 21st century. Productivity will increase 20 times. Where will this 400-fold improvement in capability come from? First, there should be a 10 times improvement in material processing capability. This is a modest assumption considering the much greater level of improvement in the past century (Ayers, 1991). Second, improvement in IT capability will increase 40 times. Although this is a stretch, high rates of IT adoption should make this possible.

For organizations, the greatest productivity improvement will be due to systematizing knowledge. This will allow great strides in improving all business processes: new product introduction, order fulfillment, customer service, and infrastructure support. In order to take advantage of these new capabilities, the coordination of large groups of people will be the greatest challenge.

Another factor influencing competition has been globalized trade. This has happened due to cheap transportation and communication. IT will only make transportation and communication better and cheaper; thus, globalized trade will continue to increase. While the direct effect on productivity will be small, the effect on competition, chiefly time based competition, will be large.

4.2 Intelligent Machines

The march towards miniaturization will continue. Miniaturization of not only integrated circuits, but also peripheral devices has allowed the prices for IT products and services to fall significantly. Miniaturization will mean the ability to mass customize machines. Basic machine technology did not improve greatly in the 20th century, and it will probably not improve greatly in the 21st century; consequently, there will only be a few new types of machinery. However, the capability of machines has dramatically increased due to computer control. The capability of machines will increase even more. There will be more application specific machines with smart sensors and intelligent, embedded devices. IT will allow much more modular functionality for machine tools which will ultimately make automation cheaper.

New automation will see 30,000-rpm machine tools, as well as machines with advanced capabilities for forming, material handling and assembly. Systems of all types will have smart embedded devices for a combination of sensing, information handling and data processing. These devices will not only be in machines for automation, but also in every type of product: cars, electronics, sports equipment, home products, etc. These products will have greatly increased capabilities based on data sensing and information processing.

In terms of new applications for information processing, embedded devices will permit better information functions inside products, better point-of-use information processing, and better adaptation of products to their environment. There will be a marriage of information processing and communications for personal telephony and business applications.

4.3 Information Technology

Perhaps the greatest improvement leading to increased productivity will be ubiquitous IT infrastructure. Improved public and private IT infrastructure, telecommunications and information systems, will allow enhanced integration of information systems. As a result, new IT applications and services will be created, which will see much more effective use of information. A greater demand for IT tools will result in mass customization of software applications. Powerful computer applications like CAD, CAE, PDMS, ERP, and EDMS, which have decreased in price from $100,000/seat in 1990 to $50,000/seat in 2000, will fall to $5,000/seat by 2010. There will be a tremendous growth in 3rd party IT based services for manufacturing and other economic sectors. IT applications and services will become commodities.

4.4 New Materials

New materials will always be developed. Plastics are the major example from the 20th century. Who knows what will happen in the 21st century? However, because of increased knowledge about materials and greater computer processing power, people will be able to order materials by specification, i.e., designer materials. This is starting to happen in plastics and should expand to metals. There will be a marriage between biologic and standard materials to create biomaterials with new properties.

The technical area with the greatest product development will be photonics. Photonics will be to the 21st century what electronics was to the 20th century. These products will not just be replacements for electronics, but will have brand new capabilities. Thus, the greatest demand for new materials will be for photonic products, and because of this demand, there will be a very great need for greatly improved methods for the processing of optical materials. The best candidate to provide this capability is nanotechnology whether this is extending the capability of MEMS (micro-electromechanical systems) or some other techniques, but miniaturization is imperative for photonics to achieve its potential.

In terms of material use, products will be designed for extended life. There will be a greater emphasis on reuse, and thus, on disassembly of products. This will be for better material reuse, and also for product enhancement by module replacement.

4.5 One Step Processes

As mentioned earlier, products will be much more complex. In order to handle this complexity, simpler design-to-manufacturing systems will be required. The emphasis will be on net shape forming. Revolutions will happen in casting, moulding, and new additive processes. For the latter, the capabilities of present rapid prototyping technologies will be expanded and become a standard means of production. Accompanying these systems will be much more capable CAD/CAE/CAM systems, which will allow more rapid design of products and processes. Also nanotechnologies will play a role in being able to create miniaturized, one-step manufactured products.

4.6 New Corporate Culture

As important as new technologies will be, the greatest imperative for corporations will be to develop both knowledge and time based infrastructure. Management of knowledge will be the lifeblood of corporations. New strategies and structures will be needed in a knowledge-oriented company. Companies will have to learn how to become more knowledge and technology based, and train staff in the new methods.

Present management methods and organizational structures are all based on managing for cost. Managing for time will shortly become a dominant strategy. The challenge will be to create time based operations and reorganize support functions in order to handle greater complexity in products and processes. Greater complexity in products will be driven by the desire for more customization, and greater complexity in processes will be driven by more globalization and shorter cycle times.

4.7 New social culture

Society will develop an information culture, which means that there will be a greater value placed on the means to create, store, move and use information. Also present culture will change such that information will be more valued than material. As a consequence, consumption of material goods will slow with the advent of information based products having a great effect on the way in which we interact with the environment.

5. FUTURE OUTCOMES

The slowdown in productivity growth due to the Information Revolution will be overcome by 2005. The industrialized world will see productivity increases return to 3 % per year as is now happening in the USA so that total productivity increases in the 21st century will be 20 fold.

There will be a dominance of IT culture in society. The 'tech' culture will become even more pervasive than it is now. New, rapid waves of IT capability will constantly be adopted by society; past examples are telephone, fax, and email. The original technology for the telephone was based on electrical circuits; fax was based on integration of electronics and telecommunications; and email was based on integration of computer technology and telecommunications. Greater integration of IT with other technologies will provide new capabilities. The infilling of IT infrastructure will integrate large segments of business and society. There will be many business opportunities for the provision of timely knowledge.

Due to the dominance of knowledge in the economy, there will be a greater emphasis on higher education. There will be requirements for higher levels of education for job entry and more training required for jobholders. The greatest demand for skills in manufacturing will be for engineering and manufacturing management; this is due to the increasing knowledge levels in product development and production and the need to manage them. In response to this requirement, industry-education partnerships will flourish in order to develop new education programs that will deliver better educated people and that will better train people already in industry.

The main manufacturing challenge will be to focus on knowledge processes in order to be competitive. The challenge is to systematize corporate knowledge, to invest in knowledge tools for higher productivity, and to manage by process in order to address time based competition. Indeed, surveys of manufacturers indicate that a major problem is the adoption of IT tools and development of IT infrastructure for better use of information (Canadian Manufacturers & Exporters, 2001; Plant, 2001).

6. SOCIAL CHALLENGES

The Information Revolution will bring many challenges to society. Societies around the world will have to build for a knowledge economy. This means changing from a 'resource or material' mind-set to a 'knowledge' mind-set. Educated people will be the resource of a developed society. Governments will have to create strategies for 'educating' all people in the society. How do we choose new directions and new areas for education as quickly as needs change, and in such a way that our society will move in the direction that the people want? Retraining people is changing from the need for upgrading skills to the need for reeducation. In addition, the public sector (governments, public services, non-government organizations (NGO's), etc.) must become more effective in providing information services for clients and in managing knowledge for society. The public sector is about half the economy of developed countries. Its structure is still based on the old paradigm of division of labour and management of cost. This must change to the management of knowledge and time in order to be able to take advantage of new opportunities to deliver better and more effective services.

The greatest social challenge, however, will be to address the widening gap between industrialized and developing countries. As the industrialized world moves more and more to a knowledge based economy, how can developing countries compete? The answer is for them to focus on education and information infrastructure. This is a daunting challenge, namely, to have a long-term education focus and at the same time to develop a production based economy to compete both in the future and in the present.

7. CONCLUSION

The Information Revolution promises great opportunities for the manufacturing sector to be more productive and effective. However, the very nature of competition is changing. Manufacturing will be transformed, as indeed will society, to knowledge and time based competition and values. The challenge is to restructure our societies so that citizens can benefit economically, and that all citizens can also enjoy their rebuilt society.

8. REFERENCES

1. Ayres RU. Computer Integrated Manufacturing. Volume I, Chapman and Hall, London, 1991.
2. Blackburn J. Ed. Time-Based Competition. Business One Irwin, Homewood, Illinois, 1991.
3. Canadian Manufacturers & Exporters. Management Issues Survey 2001. www.the-alliance.org
4. Conference Board of Canada. IT and the New Economy: The Impact of Information Technology on Labour Productivity and Growth. Ottawa, 2000
5. Drucker P. The Emerging Theory of Manufacturing. Harvard Business Review, May-Jun 1990, 94-102.
6. Drucker P. Post-Capitalist Society. Harper Business, New York, 1993.
7. Ellyard P. Ideas for the New Millennium. Melbourne University Press, 1998.
8. Haeckel S, Nolan R. Managing by Wire. Harvard Business Review, Sep-Oct 1993, 122-132.
9. Harrington J. Computer Integrated Manufacturing. Industrial Press, New York, 1973.
10. Hayashi H. A preview of the 21st century. IEEE Spectrum, Sep 1993, 82-84.
11. Hori S. Fixing Japan's White Collar Economy: A Personal View. Harvard Business Review, Nov-Dec 1993, 157-172.
12. IMS International. Proceedings of the IMS Vision 2020 Forum. Tokyo, February 2000.
13. Leaver E, Brown J. Machines Without Men. Fortune, Nov 1946, 162f.
14. Moody P, Morley R. The Technology Machine: How Manufacturing Will Work In The Year 2020. The Free Press, New York, 1999.
15. National Research Council. Commission on Engineering and Technical Systems. Visionary Manufacturing Challenges for 2020. National Academy Press, Washington, DC, 1998.
16. Plant. Plant pulse surveys 2001. www.plant.ca
17. Smith A. An Inquiry into the Nature and Causes of the Wealth of Nations. Ed. Edwin Cannan, Metheun & Co. Ltd., Strand (UK), 1961.
18. Statistics Canada. Technology-induced wage premiums in Canadian manufacturing plants during the 1980s. 1993.
19. Thurow L. Head to Head. W. Morrow & Co., New York, 1992.
20. US Department of Commerce. Higher Wages Accompany Advanced Technology. SB/93-14, 1993.

KEYWORD INDEX

MIX
Papier aus verantwortungsvollen Quellen
Paper from responsible sources
FSC® C105338

If you have any concerns about our products,
you can contact us on
ProductSafety@springernature.com

In case Publisher is established outside the EU,
the EU authorized representative is:
Springer Nature Customer Service Center GmbH
Europaplatz 3, 69115 Heidelberg, Germany

Printed by Libri Plureos GmbH
in Hamburg, Germany